Mik... ...eacon • Nik O'Dwyer

Edexcel Diploma

Engineering

Level 2 Higher

A PEARSON COMPANY

Published by Pearson Education Limited, a company incorporated in England and Wales, having its registered office at Edinburgh Gate, Harlow, Essex, CM20 2JE. Registered company number 872828

www.edexceldiplomas.co.uk

Edexcel is a registered trademark of Edexcel Limited

Text © Pearson Education Limited 2008

First published 2008

The publisher would like to thank Andy Boyce for his support in reviewing and editing the book.

12 11 10 09 08
10 9 8 7 6 5 4 3 2 1

British Library Cataloguing in Publication Data
A catalogue record for this book is available from the British Library

ISBN 978 0 435756 20 8

Edited by Sarah Christopher
Typeset by HL Studios
Original illustrations © Pearson Education Limited, 2008
Illustrated by HL Studios
Cover photos:
© iStockPhoto.com
© iStockPhoto.com
© iStockPhoto.com/Mark Evans
© iStockPhoto.com/Mark Stay
© Pearson Education Limited/Martin Sookias
Picture research by Maria Joannou
Printed in the UK by Scotprint

Websites
The websites used in this book were correct and up-to-date at the time of publication. It is essential for tutors to preview each website before using it in class so as to ensure that the URL is still accurate, relevant and appropriate. We suggest that tutors bookmark useful websites and consider enabling students to access them through the school/college intranet.

Contents

Acknowledgements

The authors and publisher would like thank the following individuals and organisations for permission to reproduce photographs and artwork:

© iStockphoto.com / Andreas Guskos; © iStockphoto.com / Nicolas Loran; © Digital Vision; © iStockphoto.com / Jillian Pond; © iStockphoto.com; © Science Photo Library / Colin Cuthbert; © Corbis / Shia Ginott; © Corbis / Sygma / Richard Melloul; © iStockphoto.com / Arno Massee; © Corbis / Bettmann; © Alamy / Metalpix; © Alamy / Imagebroker; © Science Photo Library / Tek Image; © Rex Features / Sipa Press; © Getty Images / First Light; © Pearson Education Ltd / Gareth Boden; © Pearson Education Ltd / Gareth Boden; © Pearson Education Ltd / Gareth Boden; © Pearson Education Ltd / Gareth Boden; © Pearson Education Ltd / Gareth Boden; © Alamy / vario images GmbH & Co KG; © iStockphoto.com / Stefan Witas; © Science Photo Library / Gustoimages; © Photographers Direct / Andrzej Gorzkowski Photography; © Science Photo Library / D Roberts; © iStockphoto.com / Matej Pribelsky; © Alamy / Arthur Turner; © iStockphoto.com / Alexandra Draghici; © Martyn Chillmaid; © Corbis / Yang Liu; © iStockphoto.com / Filippova Olga; © Rex Features / Nils Jorgensen; © Author; © Author; © Author; © Author; © Author; © Author; © Author; © Author; © Author; © Crown Copyright; © Author; © Author; © Author; © Author; © Author; © Author; © Author; © Author; © Author; © Author; © Author; © Author; © Author; © Author; © Author; © Alamy / Antony Nettle; © Author; © Author; © Author; © Author; © Author; © Author; © Author; © Author; © Author; © Author; © Author; © Author; © Author; © Author; © Author; © Author; © iStockphoto.com / Kimberley Deprey; © Pearson Education Ltd / Lord & Leverett; © PA / Tolga Bozoglu / AP; © Alamy / Gabe Palmer; © iStockphoto.com / Owen Price; © iStockphoto.com / Oleg Prikhodko; © Photographers Direct / Neil Holden Photography; © Rex Features / Jonathan Player; © David J Green; © David J Green; © iStockphoto.com / Vincent Voigt; © iStockphoto.com / Vincent Voigt; © Photographers Direct / Neil Holden Photography; © iStockphoto.com / Achim Prill; © Photographers Direct / Warren McConnaughie Photography; © Photographers Direct / Warren McConnaughie Photography; © iStockphoto.com / Gualberto Becerra; © Glenn McKechnie; © Glenn McKechnie; © Author; © Author; © Author; © Author; © Author; © Author; © Author; © Author; © Author; © Author; © Author; © Author; © Shutterstock / David N Madder; © Shutterstock / Girls_Vehicles; © Shutterstock / RROCIO; © Shutterstock / silver-john; © iStockphoto.com / Paul M; © iStockphoto.com / Rolf Fischer; © Pearson Education Ltd / Peter Morris; © NASA; © Getty Images / Hulton Archives; © NASA; © iStockphoto.com / Loic Bernard; © Shutterstock / KL Kohn; © Shutterstock / Les Scholz; © Shutterstock / Valery Potapova; © Shutterstock / Goran Kuzmanovski; © Shutterstock / Lezh; © NASA / Kennedy Space Centre; © Alamy / Goodshot; © Corbis / Gideon Mendel; © Shutterstock / Thomas Sztanek; © Shutterstock / Rick Lord; © Shutterstock / Christian Delbert; © Shutterstock / Sudheer Sukthan; © Corbis / Rolf Nvennerbernd / epa; © Author; © Author; © Author; © Author; © iStockPhoto.com / Mike Clarke; © Alamy / Classic Image; © Shutterstock / Joe Gough; © Shutterstock / Peter Gudella; © Shutterstock / Wrangler.

WELCOME TO THE ENGINEERING DIPLOMA

The Engineering Diploma is a ground-breaking qualification created by employers, the government and the leading education bodies to create a twenty-first century workforce. The Diploma will give you skills and experience that employers value, and will provide you with opportunities to progress on to Level 3 studies.

Get stuck in!

The Higher Diploma includes the following elements:

Principal learning The eight units covered by this book will provide you with the knowledge, understanding and skills essential to working in the engineering sector.

Generic learning Functional Skills in IT, English and maths, and Personal Learning and Thinking Skills have been embedded in this book to give you opportunities to develop and practise your skills.

The project You will have the opportunity to set your own brief, and plan, develop, deliver and review an engineering-related project. The final section of this book will give you guidance on completing this part of the diploma.

Additional/specialist learning You can choose from more than 800 different qualifications to add breadth to your principal learning.

Getting involved

As part of your Diploma, you will also need to take at least 10 days' work experience. Ideally, this should be relevant to your subject. In fact, employer engagement and applied learning are such key parts of the Diploma that you should consider which local companies may be willing to offer you work experience as early as possible to find the right placement for you.

Your school or college may already have strong links with local employers, or they may use a work-placement service. Remember that you should always speak to your teacher or tutor before contacting an employer.

Going further

The Diploma is available at Foundation (Level 1, equivalent to 5 GCSEs at grades D–G), Higher (Level 2, equivalent to 7 GCSEs at grades A*–C) and Advanced (Level 3, equivalent to 3.5 A Levels). The Advanced Diploma is recognised by universities and through the full qualification you could achieve up to 420 UCAS points.

From the Higher Diploma, you can progress to:

* Advanced Diploma
* A Levels
* BTEC National or other Level 3 vocational courses
* Work.

We hope you enjoy your studies on this cutting-edge course and that you feel inspired by the real-life scenarios and opportunities to follow your own interests in the project. Good luck!

How to use this book

This book has been divided into eight units to match the Higher Engineering Diploma qualification structure. Each unit follows the Edexcel learning outcomes and double-page (or four-page) spreads cover an individual theme or topic.

Features of the book

The features in this book are described over the next few pages.

Want to achieve more?

Each unit ends with advice on getting the best from the assessment. This tells you how you will be assessed, how your work will be marked and gives you useful reminders for key unit themes. Hints and tips give you guidance on how to aim for the higher mark band so you use your new skills and knowledge to best effect.

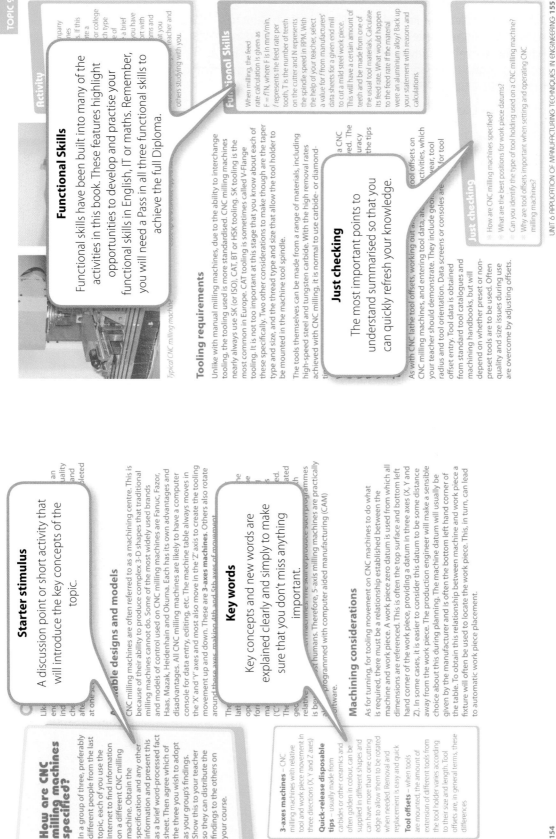

Starter stimulus

A discussion point or short activity that will introduce the key concepts of the topic.

Functional Skills

Functional skills have been built into many of the activities in this book. These features highlight opportunities to develop and practise your functional skills in English, IT or maths. Remember, you will need a Pass in all three functional skills to achieve the full Diploma.

Key words

Key concepts and new words are explained clearly and simply to make sure that you don't miss anything important.

Just checking

The most important points to understand summarised so that you can quickly refresh your knowledge.

How are CNC milling machines specified?

In a group of three, preferably different people from the last topic, each of you use the internet to find information on a different CNC milling machine. Obtain the specification and any other information and present this as a brief word-processed fact sheet. Then agree which of the three you wish to adopt as your group's findings. Show this to your teacher so they can distribute the findings to the others on your course.

Functional Skills

When milling, the feed rate calculation is given as $F = fTN$, where F is in mm/min, f represents the feed rate per tooth, T is the number of teeth on the cutter and N represents the spindle speed in RPM. With the help of your teacher, select a value for f from manufacturers' data sheets for a given end mill to cut a mild steel work piece. This will have a certain amount of teeth and be made from one of the usual tool materials. Calculate its feed rate. What would happen to the feed rate if the material were an aluminium alloy? Back up your statement with reasons and calculations.

Tooling requirements

Unlike with manual milling machines, due to the ability to interchange tooling, the tooling used is more standardised. CNC milling machines nearly always use SK (or ISO), CAT, BT or HSK tooling. SK tooling is the most common in Europe. CAT tooling is sometimes called V-Flange tooling. It is not too important at this stage that you know about each of these specifically. Two other considerations to make though are the taper type and size, and the thread type and size that allow the tool holder to be mounted in the machine tool spindle.

The tools themselves can be made from a range of materials, including high-speed steel and tungsten carbide. With the high removal rates achieved with CNC milling, it is normal to use carbide- or diamond-

Machining considerations

As for turning, for tooling movement on CNC machines to do what is required, there must be a relationship established between the machine and work piece. A work piece zero datum is used from which all dimensions are referenced. This is often the top surface and bottom left hand corner of the work piece, providing a datum in three axes (X, Y and Z). In some cases, it is easier to consider this datum to be some distance away from the work piece. The production engineer will make a sensible choice about this during planning. The machine datum will usually be given by the manufacturer and is often the bottom left hand corner of the table. To obtain this relationship between machine and work piece a fixture will often be used to locate the work piece. This, in turn, can lead to automatic work piece placement.

As with CNC lathe tool offsets, working out tool offsets on CNC milling machines, and entering tool data, are activities, which your teacher should demonstrate. They include geometry, tool radius and tool orientation. Data screens or consoles are used for offset entry. Tool data is obtained from standard tool catalogues and machining handbooks, but will depend on whether preset or non-preset tools are to be used. Often quality and size issues during use are overcome by adjusting offsets.

Key words

3-axes machines – CNC milling machines with relative tool and work piece movement in three directions (X, Y and Z axes)

Quick-release disposable tips – usually made from carbides or other ceramics and often golden in colour, can be supplied in different shapes and can have more than one cutting edge to allow them to be rotated when needed. Removal and replacement is easy and quick

Tool offsets – when tools are mounted, the amount of extension of different tools from the tool holder varies according to their size and length. Tool offsets are, in general terms, these differences

CNC milling machines are often referred to as a machining centre. This is because of their ability to produce complex 3-D shapes that traditional milling machines cannot do. Some of the most widely used brands and models of control used on CNC milling machines are Fanuc, Fazor, Haas, Mazak, Heidenhain and Okuma. Each has its own advantages and disadvantages. All CNC milling machines are likely to have a computer console for data entry, editing, etc. The machine table always moves in the 'X' and 'Y' axes and most also move in the 'Z' axis to create the tooling movement up and down. These are **3-axes machines**. Others also rotate around these axes, making 4th and 5th axes of movement.

Just checking

- How are CNC milling machines specified?
- What are the best positions for work piece datums?
- Can you identify the type of tool holding used on a CNC milling machine?
- Why are tool offsets important when setting and operating CNC milling machines?

Typical CNC milling machine

154

Engineering innovation

The ability to come up with new and revolutionary ideas for new products can lead to spectacular success in a global market. Such innovation needs a fertile and creative mind coupled with a determination to bring a new product to market. It's important that would-be entrepreneurs and inventors turn their new ideas into reality.

Knowledge-based organisation – an organisation that uses a fund of knowledge for its day-to-day operation.

Implicit knowledge – knowledge we hold in our minds but is not written down

Explicit knowledge – knowledge that exists on paper or in electronic form

The Dyson vacuum cleaner

184

What is innovation?

Innovation is about bringing something new into the world, which is better than what currently exists. It most often refers to a new product, but can also apply to services, manufacturing and management processes, or the design of an organisation. It can also include improvements to the efficiency or effectiveness of existing products, processes or services.

Innovation involves creativity. It involves taking new ideas and turning them into reality through invention, research and the development of new products and services. Successful innovation requires:

* an ability to think creatively
* a wide knowledge of existing solutions and technologies
* an ability to think unconventionally (outside the box)
* an ability to seek, apply and experiment with new possibilities and techniques
* an ability to communicate ideas and to enthuse other people
* an ability to protect and promote ideas within the legal framework.

Knowledge and innovation

Knowledge and innovation are often linked. Knowledge is usually considered to be something built up in the mind of individual people, resulting from their interaction with the world around them. However, organisations also have knowledge – within the minds of employees but also in the form of paper and electronic records. A good example of a **knowledge-based organisation** is a school, college or university. However, many large organisations and corporations now regard themselves as stores of knowledge and are beginning to realise that this fund of knowledge can be a huge asset.

Knowledge is often considered to be either implicit or explicit. **Implicit knowledge** is held in a person's mind and doesn't need to be turned into words. For example, we know that a kettle gets hot and gives off steam, so we observe caution when we approach one. **Explicit knowledge** is knowledge that has been written down, drawn, or otherwise expressed to communicate it to other people.

Case Study

Case studies show the concepts covered in this book applied to the real world through real-life scenarios. Questions and activities will encourage you to push your understanding further.

Activity

Each topic contains an activity or a short sequence of questions to test your understanding and give you opportunities to apply your knowledge and skills.

Personal Learning and Thinking Skills

Elements of the generic learning are embedded in the principal learning. These features highlight opportunities to develop and demonstrate your personal learning and thinking skills.

Case Study: The Dyson

Sir James Dyson is a well-known industr... known as the inventor of the Dyson va... worth more than £1 billion. More than ... with the performance of his convention... problem was that the dust picked up w... suction and became less efficient. Jam... the principle used in the air filter of the sp... producing the Ballbarrow (an innovative wheelbarrow that uses a ball rather than a wheel).

After five years of development in which James produced several thousand prototypes, the design for the Dyson 'G-Force' bag-less vacuum cleaner was complete. Unfortunately, no manufacturer or related distributor was willing to launch his product in the UK, because it was felt that the new invention would damage sales of existing vacuum cleaners and their replacement bags. James then turned to the overseas market, initially launching the 'G-Force' bag-less vacuum cleaner in Japan. To protect his invention, James obtained his first United States patent in 1986 (U.S. Patent 4,593,429).

In 1993, frustrated by the ongoing resistance to his idea from manufacturers, James decided to set up his own vacuum cleaner factory and research facility in Wiltshire. Dyson engineers discovered that a smaller diameter cyclone gave greater centrifugal force. This led to a way of getting 45% more suction than the original dual cyclone and removing more dust, by dividing the air into eight smaller cyclones and incorporating this principle into later bag-less cleaners. Since then, the Dyson ...

What ...

* What ...
* What ... unsatisfa...
* What re...
* What other ways are there of solving this problem?
* Where is the market for the solution and how can I reach it?
* How can I promote my idea and raise the capital I need for developing it?
* How can I protect my idea so that it does not fall into the hands of others?

Later in this unit, you will explore these in much greater detail.

...ut the Dyson ...hat is the advantage ...red with a ...wheelbarrow? ...products is the ...e Ballbarrow

...ity

Find out about the inventor of the clockwork radio, Trevor Baylis.

1 What need was Trevor trying to satisfy?

2 What did he do to satisfy that need and how well did he do it?

3 What did he use to test the principle behind his invention?

4 What event proved to be instrumental in bringing Trevor's invention into the public eye?

5 Which company first manufactured his invention?

...hat other organisations ...e involved in the design and ...facture of Trevor's invention?

...sonal learning ...d thinking skills

...or Baylis once said 'The key ...success is to risk thinking ...conventional thoughts. ...vention is the enemy of ...ress.' How did Trevor apply ...his concept to his invention? Discuss this with your group and suggest at least two other ways in which the original need could have been satisfied.

Just checking

* What is innovation and why is it important?
* What is knowledge and where is it found?
* What things does an innovator need to consider?

Introduction

It is difficult to think what our world would be like without mobile phones, computers, jumbo jets, microwave ovens, televisions, DVD/CD players or digital cameras. The engineering world has enabled us to fly, dive to the depths of the oceans, and even journey into space.

In this unit, you will discover aspects of the world of engineering. You will develop an understanding of the diverse sectors within engineering and how these interlink to offer a range of services and products. You will investigate the achievements and developments of the engineering world from a local and national perspective, and investigate the effect engineering has on the modern world.

The world of engineering offers diverse and exciting opportunities to individuals who are looking for a dynamic and productive career. You will identify the opportunities that exist within engineering, and the reward and satisfaction gained from a career as an engineer.

How you will be assessed

This unit will be assessed by your tutor who will set an assignment for you to complete. It will focus on giving you an overall understanding of the engineering world. As such, you will be assessed through an assignment that will give you opportunities to demonstrate what you know about the different engineering sectors, and their products and services, the job opportunities available within engineering, the role of professional engineering institutions, the main achievements in engineering that relate to social and economic development, and the rights and responsibilities of engineering employers and employees.

After completing this unit, you should be able to achieve the following outcomes:

1. Describe two different sectors in engineering, and explain the function and operation of different engineering products or services provided by engineering companies.

2. Review career opportunities within engineering at a local and/or national level, and the roles and functions of the engineering council and licensed professional engineering institutions.

3. Describe the key achievements in engineering in the 19th, 20th and 21st centuries, and refer to social and economic development.

4. Comment on, and explain, the rights and responsibilities of employers and employees in relation to relevant legislation.

THINKING POINTS

Many people work in engineering. Imagine what our world would be like without engineering. It will be said a lot during your studies in engineering, but just take a look around you and you should see the abundance of engineering. If you are going to work in engineering or take further courses in engineering you will need to know about a range of the different areas an engineer can work in. Think about exploring the world of engineering. When you are next on the way to college or school take a little time to make a mental note of the specialist areas of engineering around you: for example, look up into the sky, an aeroplane goes by. Yes, aeroplanes are built by engineers and their navigational systems have been developed by engineers. You perhaps saw clouds, but you expected that, because the weather forecast had said that it would be cloudy. Who developed the equipment that enables weather forecasters to predict what the weather will be like? You've guessed it; engineers.

Think about the end of your studies in engineering: do you know what qualifications you are likely to have? Do you think they will help you to get a job in engineering? Imagine being around in the 1940s and saying that you believed within 20–30 years man would be landing on the moon. How would people have taken your view? Perhaps they would think you were half crazy, but it became true because of engineers. As you learn about the pioneers of developments in engineering throughout this unit just think of the uphill battles that they had to convince other people that what they were doing would be for the good of the human race. And don't forget the legal responsibilities that engineers, and indeed their employers, need to work within. Yes, the role of the professional engineer is one that has many twists and turns but just think 'One day I could become a well recognised and praised engineer'.

Engineering as a career

Welcome – you are now on your first steps to understanding what it takes to be an engineer. Engineering is full of exciting and stimulating technologies. By following this Diploma, you will discover different job opportunities and find that the career paths are endless. Because of its diversity, engineering is one of the most challenging and rewarding roles available. Being part of it could mean that you are helping to change people's lives and the world we currently live in. You have the opportunity to shape the future!

What engineer designed this?

Look around you and pick out a few objects, from small items like a pen to large ones like a car or television. At least one engineer would have been involved in the design and manufacture of all of them. Jot down some of the different engineering skills that were probably needed and discuss your list with other students. It may surprise you to find how many different branches of engineering are involved in even the simplest products.

Sector – a grouped engineering skill or occupation

Product – a component or complete item produced within a sector, or a service carried out

Engineering is one of the most varied and challenging professions that anyone could train for. To understand what it is and what engineers do, we must first look at the different areas, also known as '**sectors**', that engineers work in.

The main sectors within engineering and manufacturing, including their **products**, are given in the table below. However, there are many more, ranging from aerospace to passenger transport engineer.

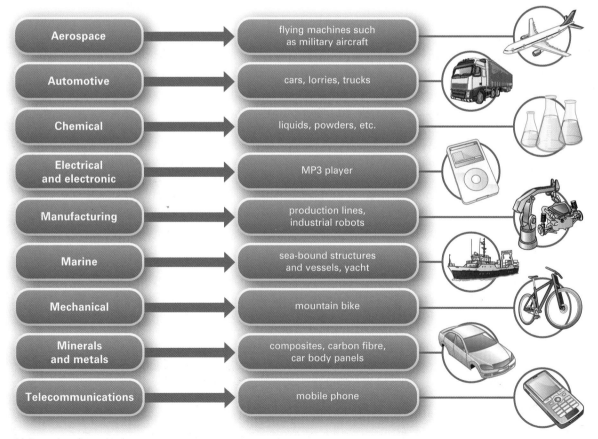

Sector	Product
Aerospace	flying machines such as military aircraft
Automotive	cars, lorries, trucks
Chemical	liquids, powders, etc.
Electrical and electronic	MP3 player
Manufacturing	production lines, industrial robots
Marine	sea-bound structures and vessels, yacht
Mechanical	mountain bike
Minerals and metals	composites, carbon fibre, car body panels
Telecommunications	mobile phone

Main engineering sectors

Products in the mechanical sector

Products in the electrical/electronic sector

As you can see from the images, the types of products within a sector are wide-ranging. However, they all have one thing in common: they have all been engineered in some way.

From this task you should have discovered that a product can be the result of a number of sectors. Although knowing the sector is important, it is understanding **why** that counts most. The reasoning behind all of your judgements is important and, as an engineer, you must have a clear understanding of what and why.

Personal learning and thinking skills

Think about a mobile telephone. What sector does this come from? Be careful: products that have many components may come from more than one sector. Write down your answer, including a reason. One sector could be *'Electrical/electronic, due to the circuitry inside which makes it function correctly.'* Your teacher will need both your answer and reasons to prove you fully understand.

Activity

You can do this by looking through the local press or internet. Select at least four companies and produce a report to include the following:

* company name
* address
* sector for which they produce products
* the products themselves.

As an extension, show an example of a product and justify the sector or sectors to which it belongs. It might be useful to annotate your images.

Just checking

* Engineering is broken down into what?
* Is it true that a product can only come from one engineering sector?
* What is the most important thing when placing a product into a sector?

Identifying engineering sectors

How many sectors make a plane?

Imagine you are to design a new passenger aeroplane for British Airways. Inside the plane will be the latest technology to control it while in the air. The passengers will have access to the latest entertainment available, and the fixtures of the plane will all be modern. How many sectors do you think will need to be involved in this creation?

From Topic 1 we know that all engineering can be divided into sectors. But how do we know which products fit into which sectors? Often a **component** could fit into several. To understand this fully, you must first have some understanding of the operations and processes carried out to make a product.

Stages in production of any product

However simple a product, it will have gone through at least the following stages:

* defining the concept, to meet a need or a gap in the market
* a series of design stages, until the product is understood in detail
* development of a prototype
* testing and further development
* manufacture.

This is a very simplified list. Some of these processes may overlap or be repeated before a lot of money is committed to buying components and starting large-scale manufacture. You will go into these stages in more detail in other units. However, all these processes require engineers with different skills working together.

The aerospace industry

The aerospace industry employs a high proportion of engineers worldwide. Look back at the diagram in Topic 1, which lists the main engineering sectors. You can see that engineers from all these sectors are required by the industry, except possibly automotive and marine.

To identify all the sectors involved in making a standard commercial aeroplane, we need to think about its main parts and see which sectors these fall into. Obviously, we can list 'aerospace', as this is where the final product is going to be used. The entertainment system within the plane would come under electrical and electronic, but so would the navigational system. The structure of the plane requires highly skilled mechanical and structural engineers, who are also exploiting developments in new materials. The shell of the plane is also mechanical, but it will also need specialists in coating materials.

Three sectors, with possible definitions, are given in the table.

Components – parts that go together to make a product

Cross-sectional view – a view showing the inside of a product as if it has been cut and opened out

Sector	Definition
Aerospace	Design, development, construction and testing of all flight vehicles
Electrical/ electronic	Design, development, construction and testing of all electrical and electronic components, circuits and systems
Mechanical	Design, development, construction and testing of all mechanical components, structures and systems

Activity

Look at the definitions for the three sectors in the table and the cross-sectional view of a passenger aeroplane. All of the plane's parts utilise the expertise of engineers working within different sectors. In pairs, list as many parts as you can think of in our 'New passenger plane' that fall into each of the three sectors. You may come across a number of parts that fall into more than one sector, so remember to explain why this is the case.

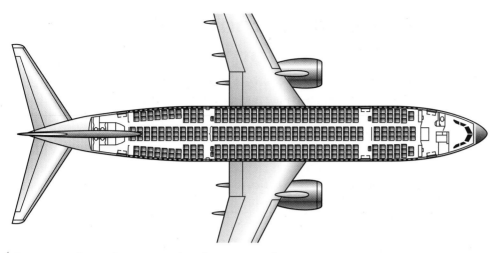

Cross-sectional view of a passenger plane, from nose to tail

The electrical and electronic industry

Think how many objects sold today are electrically powered. Even for those that are not, computers are involved in their design and electrically powered machines in their manufacture. The roles of electrical engineers cover everything from nanotechnology, where electron microscopes are required to 'see' developments, to massive electrical generators, motors and transmission systems.

As with the aerospace industry, there is wide overlap in the engineering skills required to support any major development. In the electrical power industry, there is much emphasis now on developing 'green' sources of energy. Wind farms exploit the knowledge of aeronautical engineers to design the turbine blades, mechanical and electrical engineers to make them rotate, civil engineers to construct the towers and even marine engineers for off-shore wind farms.

The need for different engineering sectors to work together can even be seen in a fairly simple electronic product, like a computer mouse.

This uses a mechanical ball to register movement of the cursor. The ball's movement is interpreted by a series of rollers. These rollers convert the physical movement into an electrical signal, which is then sent to the computer and the cursor makes the appropriate action on screen.

Ball-style mouse, showing components

Activity

Using the example of the computer mouse, produce a word-processed document or a PowerPoint™ presentation on how a computer mouse works. Start by producing a table for all the components, then highlight the engineering sectors involved in their design and production.

You could also explain the purpose of each part you have highlighted. Remember to justify your answers, giving clear reasoning for your decisions.

Just checking

* What sector would you put flying vehicles into?
* Why do most products fit into more than one sector?
* What kind of signal is the movement of a computer mouse turned into?

What must you have next?

With technology fast changing the way we live, and new products entering the market every week, what do you think will be the next 'must-have' item? Think of everything that you own at the moment. In recent years, video technology has been integrated into mobile phones and Ipods. What more features could they have and how could they be better? Do most people use all the latest improvements or are they more to attract new customers, who must have the latest of everything?

Composite – a mixture of materials, put together to achieve certain characteristics

SatNav – satellite navigational system exploiting GPS (global positioning system)

Commercial product – a product that meets a market need at an acceptable price

Annotated – labelled, written information on an image

Adoption of new materials and technology

Few industries would survive for long in any of the engineering sectors if they did not embrace new materials and technology. Some sectors are faster moving than others, and depend on frequently upgrading their products to stay ahead of their competitors. An obvious example is the mobile phone industry, which comes out with improved or new products all the time. However, even bicycles are not immune from this: many now use light and strong **composite** materials, which can put their cost up to several thousand pounds each.

Utilising new technology

So far you have looked at what products go into different sectors. To understand how sectors utilise the technology available, you need to understand how products work and what problems they can encounter.

Look at one of the latest technologies to start entering the mass market, satellite navigational systems (**SatNav**). They were almost unheard of five years ago but are rapidly becoming a necessity, particularly for commercial drivers who need to deliver items to new addresses on time. They are already the subject of 'add-on' features. However, SatNavs are not free from problems, as well-publicised incidents, such as large vehicles being directed into narrow lanes, show.

Where new technology comes from

Much of the advanced technology that we use today initially started its life as part of a military experiment, or as a spin-off from the space race. Then, as the technology became well known and the cost came down, it found its way into the **commercial** market. This was because 'blue skies' research into new materials and technology was either too expensive or too much of a gamble for the commercial world.

Car navigational system

A major shift has occurred in the last 10 to 15 years in the development of new materials and technology. A lot still comes as spin-off from military and space developments, but the greatest advances are probably now coming from civil developments for mass markets, particularly in the exploitation of computer and electronic advances.

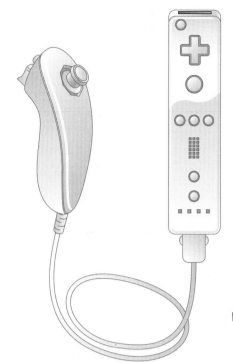

Wii controller and nunchuck attachment

The mobile phone is a good example. To stay ahead of competitors manufacturers are continually striving to make these lighter, smaller, more capable and able to operate for longer without recharging. The profits in this market are paying for research and development of new materials, new battery technology and new manufacturing techniques, which are then being exploited by other industries. The military are already finding it more cost-effective to monitor commercial developments and adapt them! Environmental concerns are also resulting in new materials, technology and manufacturing techniques for vehicles and aircraft.

Activity

In this unit you will need to be able to explain in detail not only products from each given sector, but also how they work and how the technology is utilised. Imagine that you are a reporter for an engineering magazine. It is your job to find out about two new products (your teacher will help you decide on which ones). They could be from the same sector or different sectors. You are to report on the products stating the following:

* what the product is and what sector it comes from
* how the product performs its designed function
* the type of technology used in the most up-to-date model
* a comparison with the first commercially available model.

As an extension, you could highlight any areas you think could be developed further, stating why and how.

Remember to include images of the products and ensure they are **annotated** fully.

Just checking

* What are some of the driving forces that cause new materials and technology to be developed?
* How important are military requirements in the development of new technology?
* What are the advantages and disadvantages of SatNav systems in vehicles?

A career in engineering

A job for life; a career where you continually train, develop and grow; opportunities locally, nationally and internationally… engineering has them all!

Engineering companies employ people in job roles that require different skills and academic qualifications. Two particular categories of job role are those of engineer and engineering technician.

To be a good engineer requires a combination of high academic achievement, practical knowledge and excellent interpersonal skills, particularly because they will be working in a senior role in a company. Typically, they work on the design of new products and will have contact with customers, company managers, people who manage machines and assembly departments, and engineering technicians. They make decisions and carry out actions that are of a strategic nature.

Engineering technicians are people who hold qualifications specific to a particular type of job, for example setting up computer-controlled machinery. They will be expert in what they do, make decisions and report back to an engineer.

After leaving school, qualifications can be taken full-time or part-time. The part-time route takes longer, but you will be earning as you learn, and also gaining valuable experience of working with people.

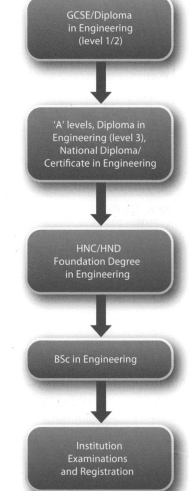

What career do you want and where?

Imagine that you have just left university with an engineering degree. Where could you work? What type of engineering would you like to do? How much could you be paid? In pairs, look into the future (as this is what real engineers do every day), and write a short description of your ideal engineering job, the initial pay you could expect and your likely daily work. What about your location: where in the world would you like to be?

Case Study

DS left school at 16-years-old with eight good GCSEs and was all set to start on 'A' levels at the local FE College. However, as a result of work experience, she decided to take up an offer of an apprenticeship with the company. She went to college one day per week and over five years achieved a National Certificate, and a Foundation Degree in Mechanical Engineering. On completing her apprenticeship she was employed as a CAD design technician. DS then studied part-time for an Honours Degree and promotion to senior product designer. DS is now responsible for the design of new products, supervising CAD technicians and negotiating the technical aspects of contracts with customers. Soon, she will have enough experience to be able to apply for membership to one of the engineering institutions. DS is now 26-years-old.

Routes into engineering

Not all engineering employment starts with a degree. There are many levels of engineering, offering different levels of commitment, training, travel and expertise. How you succeed in your chosen area will relate directly to the amount of continual training you undergo and skills you acquire. Since 2004, it has been a requirement for schools and colleges to provide experience of work through work-related learning (**WRL**) – you may have been lucky enough to have gained from this. It should give all students in the future practical experience and understanding of work in many areas, including engineering.

For most engineering routes, you will need to have achieved good maths and English qualifications, so bear this in mind with your other studies. But, assuming that you feel ready to start your engineering career after you have completed this Diploma, what opportunities are out there?

Some apprenticeship routes are listed in this table:

Age	Qualification	Description
Pre-16	Young apprenticeship	50 days work-related learning, studying usually at college while still at school to complete the **PEO** (Performing Engineering Operations) **NVQ** Level 2
Post-16	Young apprenticeship	Full-time at college Part-time while in employment

Exploring career opportunities

Imagine you have been in your first engineering role for three years and now feel ready for a change. You want to progress and diversify, but what career route will you take? How will you manage the move? What opportunities are there for you once you are in your next job? Engineering is advancing all the time and engineers must be ready to change with it.

Once in engineering employment, there are many routes you can follow. Generally, these depend on your interests, but also your skills and ability to think and understand. Research and development (R & D) engineering is often considered the cutting edge of the profession. Within this area you will be expected to challenge everything – and design and develop tests to make things fail, so that you can come up with solutions to prevent this and generate improved products for the future. However, R & D is just one area within engineering. You can also apply research to seek out and develop your own career opportunities!

> **WRL** – Work-related learning, which implies that the learning is achieved through an employment-style situation
>
> **Pre-16** – under the age of 16
>
> **Post-16** – over the age of 16
>
> **PEO** – Performing Engineering Operations, an industry-related qualification
>
> **NVQ** – National Vocational Qualification

Activity

Career path portfolios

As an extension to the earlier task, you are to produce two career path portfolios to be used as career guides: one for an apprentice engineer, the other for a graduate engineer, both in the same engineering sector. You may need to use the internet to research relevant case studies for engineering careers. You could start with the following website: http://targetjobs.co.uk/engineering/

You will need to invent each person, giving them a background, and explaining where they come from, what activities they enjoy doing, and how this relates to their chosen profession.

Each portfolio should contain at least the following information:

* age at which employment could start and a possible age for reaching each new career stage
* qualifications and/or skills needed to enter each stage
* a progression route.

The two engineers can end up in different positions. However, the career path portfolio must explain exactly how they achieved their progression. As an extension, include information on typical wages/salaries that could be obtained today for each employment level. You will find information in websites listing engineering appointments. Also trade magazines like Electronics Weekly frequently include surveys of salaries within their own sector.

Just checking

* In what two subjects will you need good results, if you are to follow an engineering career?
* How many days of WRL does the PEO offer?
* Is an apprenticeship only available to post-16 learners?
* Do engineers who join a company as graduates always end up with better jobs than those who enter by other routes? If so, why? If not, why not?

The role of professional engineering institutions

Why there is a need for Engineering institutions and societies

Engineers have a duty to act in the best interests of the general public because everyone uses the products and services that they design and manufacture. If a product fails in service and this causes death or injury then the company that made it can be prosecuted for criminal negligence and sued for damages.

To prevent this type of scenario happening, companies must employ staff that are properly educated and trained – particularly those who hold senior positions and are responsible for making big decisions.

For example, when we travel in an aeroplane at an altitude of 10,000 metres how do we know that the cabin designer calculated the thickness of its skin correctly? Someone will have signed a document confirming that everything is as it should be: this is a huge responsibility and something that can only be done by a licensed engineer.

The Engineering institutions and societies are there to set and monitor the benchmarks for engineers becoming licensed in the various disciplines of engineering. Another hugely important job they do is to promote the positive role of science, engineering and technology throughout the world, and to encourage talented young people to join the profession.

The Engineering Council UK

EC^{UK} is an organisation set up by Royal Charter to regulate the engineering profession in the UK. It publishes a brochure called *Regulating the engineering profession*, which you should read through to find out why regulation is needed and how it is carried out. You can access the brochure on line at http://www.engc.org.uk/documents/ECUK%20Brochure.pdf.

The EC^{UK} states that it is resolute in its commitment to the maintenance of high standards within the engineering profession, coupled with a belief that entry to the profession should be made available to as wide a range of people as possible. It works with universities, industry and the institutions to develop routes to registration that integrate education with supervised work-based professional development. You may be interested to know that EC^{UK} has participated in the development of the Engineering Diploma you are currently studying so as to ensure that what you learn is relevant and will provide an excellent foundation for your career in engineering.

To find out even more about the EC^{UK}, go to its main website, http://www.engc.org.uk/about_ecuk/default.aspx. You could also talk to a Chartered Engineer about EC^{UK} when you are on work experience.

The Professional Institutions and Societies

Professional Institutions and societies, for example the Institution of Mechanical Engineers (I Mech E) and the Royal Aeronautical Society (RAeS) represent the interests of specific types of engineering. The oldest ones were set up in the mid 1800's and their purpose was, and still is to promote professionalism in engineering. The first president of the I Mech E was George Stephenson who is famous for being an excellent designer and builder of railways.

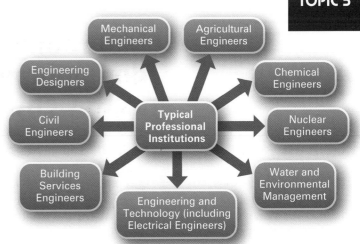

Before you can apply for registration with the EC^{UK} you must first become a member of an EC^{UK} licensed institution or society and to do this you will need to pass examinations and be able to prove that you are a competent engineer. There are different types of membership grade and they range from affiliate/student through to full member and fellow. All of the institutions and societies have very good websites which explain what you would need to do to gain membership.

Career opportunities

As you read at the beginning of this unit, engineering is one of the most challenging and interesting professions that anyone can train for. Unfortunately, in the UK the title 'engineer' is not protected, which means that people without any proper training can call themselves engineers. This gives the profession a bad name and is one of the reasons why the EC^{UK} and the Institutions are so involved with training, career progression and the maintenance of standards.

Engineers at work

Obtaining a qualification at college is just one step on the path to becoming a fully functioning engineer. When you start work in industry, you will be expected to carry out continuing professional development (CPD). How you do this is up to you and the company that you work for, but if you wish to develop your career fully, this will involve more training and examination, to a standard which meets the requirements of an institution relevant to your sector of engineering.

Activity

Use this link to find out more about an institution: http://www.theiet.org/

✱ What does the vision statement say?

✱ In how many countries does it have members?

✱ To join as a student member what must you be doing?

✱ To become a fellow what does an engineer have to prove.

Royal Charter – granted to an organisation to ensure that it is legitimate and regulated in the way it functions

Just checking

✱ Why do engineering institutions and societies exist?

✱ What is the role of EC^{UK}? How do you become a member?

A nuclear power station

Household item

Major engineering developments in the 19th Century

What has engineering done for you? In fact, what has engineering done for us all? By now you should have started to understand that the majority of things with which we interact in our everyday lives have had the input of an engineer, in one way or another. This includes creating and adopting new materials and technology, creating new products and developing new systems. But what have been the main achievements and how have they affected the **social and economic development** of society?

What drives new developments?

Engineering is all about making products to simplify and enhance life. Less than 100 years ago, most houses in the UK did not have mains electricity, mains water or sewage. Some still don't in remote areas, particularly mains sewage. Most inventions, which we now take for granted, would not be possible without earlier discoveries. Some of these may have occurred by chance, but still depended on someone with an engineering mind to recognise the potential of what had been observed. Others have only come from persistent trial and error, with individuals and teams refusing to be beaten by repeated failure, until eventually there was a successful outcome.

The development of electrical power

One of the most significant developments of the 19th Century was electrical power. However, some of the effects of electricity, particularly **static electricity**, had been noted by the early Greeks 2000 years before. From early in the 17th Century, scientists and engineers were experimenting with these effects. However, it wasn't until 1827 that electricity was quantified by Georg Ohm. In 1831, Michael Faraday developed the early **dynamo**, generating an electrical charge using a magnet. This was to be the start of electrical generators and probably the first practical use of electricity. It was not until 1878 that the first public power station opened in London, followed later the same year in New York by the work of Thomas Edison. Edison's first efforts provided lighting at 110 volts direct current (d.c.) to just 59 properties. We have come a long way since then. Interestingly, the USA still provides single phase power at 110 volts, though with alternating, not direct, current.

Activity

In no more than 150 words for each, produce a report on three items that were around in some form in the 19th Century and are still used today. Examples are the kettle, cooker, vacuum flask and electric lamp. Your report should highlight the most up-to-date versions of each, showing the features and all of its functions, together with the materials used. This information should be used as a comparison with the earliest model you can find (for this, again, highlight the features and materials used).

As an extension, include the original inventor or inventors, the date of the invention and, if possible, the cost.

Other developments in the 19th Century

Electricity was just one area of development in Victorian Britain, which was undergoing major social and economic changes. The country was producing vast amounts of exports to its empire and collecting raw materials in return. Civil engineering was making a large contribution to the growth of industrial cities throughout Britain, with rapid improvement of public and commercial transport. Most of the canal systems installed in the previous two centuries were superseded by a vast and complex railway network. This provided cheap travel for working class people and encouraged the development of holiday resorts and leisure activities. Another key development, driven by this rapidly developing infrastructure, was the standardisation of time.

Prior to this, there was a time difference of 20 minutes between the east and west extremities of Britain, because people marked noon (midday) when the sun was directly overhead. Standardisation of time was called Greenwich Mean Time (GMTs) – a system still used today.

These advances were being paralleled in other countries, particularly in Europe and the USA. There was intense competition leading to rapid industrialisation, which resulted in rapid engineering developments in many fields. The Great Exhibition in 1851 in the Crystal Palace, London was a showpiece for the achievements in the world at that date. Most of these developments improved the lives of some of the population, but the social and economic effects often created a worse environment for the majority, especially in the new industrial towns. As ever, war and the threat of war was a driving force behind many advances.

Social and economic development – how people's lives change, including their wealth

19th Century – 1800 to 1899

Static electricity – a charge that can build up due to friction between two surfaces

Dynamo – another name for an electrical generator, a machine that converts mechanical into electrical energy

Activity

As an engineer, you should know something of the great engineering pioneers. We have already mentioned Ohm, Faraday and Edison in the electrical field. You have probably heard of Isambard Kingdom Brunel, responsible for structures such as the Clifton Suspension Bridge and the Great Western Railway.

Select one of these, or another great engineer of the 19th Century. Using the internet or library, produce a short biography of no more than 500 words. Some questions you might consider are:

* Where did they come from?
* How did they become engineers?
* What were their major contributions?
* Was there any opposition to what they were trying to do and did they have any failures?
* Are we still benefiting from their advances?

Personal learning and thinking skills

The 19th Century produced major achievements in manufacturing and the development of new technologies. It also saw a rapid decline in other industries, which led to widespread poverty.

Use the internet or library and see if you can identify at least three key activities that gave people work in the 18th Century, which declined drastically in the 19th Century. What were the main effects? What happened to the people who were affected? How easy was it for them to adapt to new developments and industries? A good starting point might be to research 'cottage industries' and 'agriculture'.

Just checking

* Who quantified electricity in 1827?
* How many properties were supplied in the first electrical network in the USA?
* When was the Great Exhibition and where was it held?

Engineering in the 20th Century

As an engineer you should be constantly questioning why things happen the way they do, and how they could be improved. The 20th Century saw a rapid acceleration in the development and exploitation of new materials and technologies. Why did it happen? What impact have the changes made?

The breadth of 20th Century development

It is impossible even to list the major engineering developments of the 20th Century in a short topic like this. Just think of advances in transportation (rail, road, sea, air and space), civil engineering, communications, computing… the list is endless. Instead we will pick out a few key areas to explore in a little more depth.

What drove this development?

Why were there such rapid advances in such a short time, compared with preceding centuries? It is an unfortunate truth that war, or the threat of war, has been the biggest driving force. The build up to the First World War saw major heavy engineering advances, especially in ship building. Thousands of weapons and millions of rounds of ammunition were produced, resulting in advances in mass production techniques. The aircraft was little more than an expensive toy for enthusiasts in 1913. By 1919, there were fighters, bombers and transport aircraft, leading to rapid civil aviation development in following years. There was a similar surge during World War II.

During war, engineering processes are compressed, sometimes with high levels of risk, which are justified because the results are needed. Also, governments are prepared to invest vast sums in R&D to counter threats to their countries. World War II resulted in advances like radar and the atom bomb.

The space race was part of this too. Yuri Gagarin became the first man in space in 1961, followed eight years later by the successful moon landing led by Neil Armstrong. Today orbiting satellites give almost instantaneous communications anywhere in the world.

Computing technology now enables rapid prototyping, and mass markets for consumer electronics products provide civil money for R&D. Demands for medicines and concerns for the environment may also have a bigger impact than defence on future advances.

The private car

Cars were around in the late 19th Century, but only for rich enthusiasts. This changed in 1908 with the Model T, manufactured by the Ford Motor Company, USA. It saw the first fully **automated** production lines: assembly in 98 minutes from start to finish.

One small step for man; one giant step for mankind

The motor that moved the world

Automated – carried out partially or completely by a computer or machine

Alkaline storage battery – a type of battery or cell, dependant upon the reaction between zinc and manganese

Activity

Fifteen million Model T Fords were produced. They came in only one colour: black. The original selling price was $800, but by the end it was down to just $350. In no more than 150 words, explain the following:

❋ Why do you think they stopped manufacture?

❋ Why do you think that it only came in one colour?

❋ What is the advantage of an automated production line?

❋ Why was it possible for the cost to be dropped by so much?

Why was the automobile so life-changing? Due to the price, personal travel was possible. Instead of living in a city, you could commute from the smaller towns, villages and the countryside. This helped to bring money into these areas and with money came bigger shops, better facilities and growing communities.

Communications and development of batteries

The telegraph had been well developed by the end of the 19th Century. However in 1901, Marconi successfully managed a transatlantic transmission of a message. The radio was soon developed by the military, but it also provided mass entertainment and worldwide communication. The world was starting to become a smaller place.

Also in 1901, the **alkaline storage battery** was invented by Thomas Edison. It provided portable electricity for communications, in vehicles and even just to provide torches to move around in the dark.

Activity

If you could be an inventor in the 20th Century, what would you most like to have invented? How would it work and why would people want it? Choose any of the ideas above or research other inventions and discoveries. Produce an advert for your product, explaining what it is, who it is for, what it does and how.

As an extension, imagine that your invention is on sale. Has it been a success? What impact has it made? Produce a report that reviews these questions, and end with a suggestion on an improvement for your design.

Just checking

❋ In what year was the Model T first manufactured?

❋ How long did assembly of the Model T take?

❋ What was the total number of Model T Fords manufactured in black?

❋ What two things did Marconi achieve in 1901?

❋ In what year did Neil Armstrong walk on the moon?

Where is engineering going in the 21st Century?

Is there a direct relationship between technological developments and the way we are able to live our lives? This depends on the type of technology, and the way in which we connect with it. Take radioactive energy: for some of us, our electricity is generated using nuclear power stations, but as individuals we don't have our own power plants in our back gardens.

Achievement and development

Engineering achievement and development are different, but one cannot happen without the other.

We should all be able to list major engineering achievements that have occurred within the last two centuries, but what have been the developments behind the achievements to make them happen?

As an example, look at the moon landing. This was made possible through major improvement in rocket design (which was heavily based on developments by Nazi Germany of the V1 and V2 rockets to strike London) and advances in rocket fuel. It also could not have been achieved without computer developments to perform the complex calculations necessary to navigate to the moon, and safely re-enter Earth's atmosphere. The space programme also required new materials that are now routinely used in common products – often called 'spin-off'.

Activity

Working in groups of two or three, select five major products or developments of the 20th Century that may have created, or are likely to create, problems in the future. For each one, list its advantages but also the possible problems. Then try to suggest practical solutions for reducing the latter without losing too many of the advantages. You might consider the private car, aviation, communications or even packaging; also impacts such as health risks through pollution, noise, destruction of the environment and safe disposal.

Produce a PowerPoint presentation of this lasting no longer than six minutes.

What further engineering advances can we expect?

This is an exciting time to be starting a career in engineering. The biggest problem may be what area to choose to go into. Computing power is developing at a phenomenal rate, at the same time as size and price are reducing. There is probably more computing capability in a modern digital camera than in most so-called 'super-computers' of 20 years ago. Production of ships, aircraft, vehicles and heavy machinery still requires heavy engineering skills, but robotics is rapidly changing the way many products are manufactured.

The World Wide Web enables scientists and engineers to exchange ideas, but also to find out about new materials and technologies, being developed for very different purposes. Computer simulation is already so capable that complex engineering projects can be taken to advanced stages of design, and even testing, without any metal being cut.

Advances in technology have involved extensive testing – including sending chimps into space.

Super miniaturisation via nanotechnology is promising revolutionary new materials, and also major advances in the ability to inspect the performance of machines. This can help to optimise performance but also avoid catastrophic failure. Its use for monitoring, and even repairing, the human body is being actively researched.

In fact, the boundaries between engineering sectors are becoming increasingly blurred. This was recognised in 2006, with the merging of the Institution of Electrical Engineers and the Institute of Incorporated Engineers into a new institute called the Institution of Engineering Technology (IET). Combining technologies also tends to produce new engineering sectors, as a group of engineers develop new skills in the combined areas.

High-speed rail

Activity

Look at the latest mobile phones. Now compare these with mobile phones of only 10 years ago. In groups of two or three, list as many new features on the latest phones as you can find, including materials, battery power and construction. Produce these as a column in a spreadsheet. In the next column, write the new developments in engineering that have made this possible.

As an extension, suggest further desirable improvements. Divide these into those that should be achievable with current technology and those that would need new developments.

Activity

Travel plays a big part in people's lives, for employment or for pleasure. Many imported goods are also distributed using the rail network.

With the planned introduction of high-speed rail across the UK during the 21st Century, what will be the impact on economic and social life?

Imagine that you are a spokesperson for the company responsible for running the high-speed network. Produce a press release that highlights the positive economic and social effects of this achievement. You could start by explaining that time to travel across the country will be halved, explaining why this is positive.

As an extension, you could put yourself in the position of someone who is against the high-speed rail link. Highlight the negative economic and social effects that could result, making sure you justify the points fully.

Achievement – a milestone in history
Development – an action, invention or service that aids in making an achievement

Just checking

* What changes could happen to engineering sectors in the future?
* What does economic impact mean?
* What does social impact mean?

Rights and responsibilities of employers and employees

You have probably heard about the **Industrial Revolution**, when workers – including children – worked long hours in cramped conditions for inadequate wages. Whether we work in a factory, in an office or at home, our working lives are now governed by laws passed by Parliament. These are designed to protect the workforce from excessive demands by employers. They also protect the employer against wrongful acts by their employees. In other words, they place obligations on employers to treat their employees fairly, but also on employees to play fair with their employers and other employees.

Employment legislation

In the UK, the principal rights and obligations imposed on employers and employees come from:

* common law, covering contracts of employment between employer and employee, based on what has become acceptable in the past and decisions made in the courts

* UK legislation

* European legislation, which is then adopted by the UK Parliament.

Some rules and regulations are **statutory**, which means they have been passed by an Act of Parliament and it is a criminal offence to disobey them. In addition, there are Regulations which, although not statutory, amplify the law.

Unless specifically stated otherwise, employers, employees and the **self-employed** should always abide by them. The Department for Business Enterprise and Regulatory Reforms (previously the DTI) is responsible for implementing employment legislation. Its roles are given on its website http://www.berr.gov.uk. In particular, it is the Government ministry with lead responsibility for championing the interests of employees.

Employment Rights Act 1996 and the Employment Act 2002

The Employment Rights Act 1996 was amended by the Employment Act 2002. It covers all areas of employment, including statutory minimum rights as follows:

statement of employment (a contract)	statement of pay (wage slip including deductions)
no unauthorised deductions from pay	minimum wage (reviewed and updated regularly)
minimum holidays	disciplinary procedures
maximum working hours (with exemption by agreement)	access to an industrial tribunal to resolve disputes
maternity/paternity/adoption leave and pay	redundancy pay.

Health and Safety at Work Act 1974 (HASAWA)

How health and safety is implemented in the work place is covered
in more detail in Unit 4. The legal requirements are laid down in the
HASAWA and are enforced by the Health and Safety Commission (HSC)
and the Health and Safety Executive (HSE). The Act regulates all aspects
of risks to the health and safety of people at work. The HSE enforces the
law, with powers to enter and seize items, take photos, statements and
issue enforcement notices or even close a work place down. The Act is
amplified by several other pieces of legislation, including these:

Health Act 2006, which, among other things, covers smoking in the workplace and in vehicles	Workplace (Health, Safety and Welfare) Regulations 1992
Management of Health and Safety at Work Regulations 1999	Control of Substances Hazardous to Health Regulations 2002 (COSHH)
Health and Safety (First Aid) Regulations 1981	Noise at Work Regulations 2005
Control of Major Accidents and Hazards Regulations 1999 (COMAH) amended 2005	Manual Handling Operations Regulations 1992
Health and Safety Information for Employees Regulations 1989	Provision and Use of Work Equipment Regulations 1998 (PUWER)

Remember that, although your employer is responsible for ensuring
that you can work safely, as an employee you are equally responsible
for obeying the rules and not putting yourself, other employees or
any members of the public, at risk by your actions. This is particularly
important in engineering, as many of the activities you will be involved in
involve risks, and you should always be attempting to minimise these.

Other Acts and Regulations affecting employers and employees

Equality Act 2006

This Act has amended or superseded several other Acts covering relationships between employees, as well as between employers and employees. In particular it established the Commission for Equality and Human Rights (CEHR). It makes it unlawful to discriminate on the grounds of religion or belief in the provision of goods, facilities and services, the disposal and management of premises, education and the exercise of public functions. It also requires public authorities to promote equal opportunities between men and women, and prohibits sex discrimination when carrying out their functions.

Employment Equality (Age) Regulations 2006

These Regulations cover discrimination in employment on the basis of age. They also lay down rules to help prevent victimisation or harassment of employees by employers or other employees.

Disability Discrimination Act 1995

This Act gives rights in work to people with disabilities. Employers cannot discriminate between able-bodied and disabled applicants for a job. All employees should be selected on merit. It also affects other things inside and outside work, such as access to goods, facilities, services, buildings and transport.

Employer – a person or organisation that employs one or more persons under a contract of employment

Employee – a person employed by an employer under a contract of employment

Industrial Revolution – rapid development during the 19th Century, which saw a major shift of people's work from agriculture and traditional crafts to towns and factories

Statutory – when something is binding in law and it is a criminal act not to obey it

Self-employed – anyone who works for gain or reward other than under a contract of employment

Race Relations Act 1976 amended 2000

This Act makes it illegal to discriminate on racial grounds for training, employment, the provision of goods, facilities and services and other specified activities. Employers can also be held responsible for acts of racial discrimination committed by employees, unless they have taken all reasonable steps to prevent this.

Sex Discrimination Act 1975 amended 2003

Discrimination on the grounds of gender and/or marital status is unlawful in employment, education and training. This is enforced by the Equal Opportunities Commission (EOC). The amendment in 2003 extended it to the police.

Data Protection Act 1998 and safeguarding of other information

The Data Protection Act prohibits personal information held by an individual or organisation being shared with others without permission. It also applies to employees passing on personal information about others that was obtained in the workplace, and their personal details being used outside work. Data must be fairly and lawfully processed, used for limited purposes, relevant, accurate, not kept longer than necessary, held securely and not transferred to other countries without adequate protection.

Companies succeed because they have good products and robust marketing plans. Many companies will require you, as an employee, to sign an NDA (non-disclosure agreement), forbidding confidential information being passed to anyone outside the company without permission, or taking information with you if you leave the company for any reason. However, even without a formal NDA, all employees must be careful to safeguard information that they gain at work and not pass it to anyone who should not receive it. This is your obligation as an employee.

Activity

The Equality Act has only recently come into law. It includes many provisions previously covered by some of the other Acts listed above. Working in groups of three, use the internet to look at this Act and identify the main provisions that affect you as employees, especially those that are new within the Act. Then divide between you the other five Acts and Regulations that follow it above. Look these up as well and see which of their provisions affecting employees have been included in the new Act. Identify any that have not. Make a checklist of the key points and compare your list with those produced by other groups.

Just checking

* What new legislation affecting employment came into effect in 2006?
* What is the difference between an Act and a Regulation?
* Who is responsible for ensuring that the workplace is a safe place to work?
* What Act or Regulation covers smoking in the workplace?
* In what circumstances might you be required to sign a non-disclosure agreement?

Unit 1 Assessment Guide

In this unit you will investigate the different categories of engineering found in the UK and discover examples of manufactured products and services provided by companies. You will also gain an insight into careers in engineering and what types of qualifications and training are needed to enable people to work effectively in the industry.

Engineers are important because they develop products which improve the quality of life and also generate income for the country. To ensure fairness in business there are rules, in the form of legislation, which have to be complied with by both the employer and the employee.

Time Management

Manage your time well as this unit has a number of different components that will have to be researched. Ensure that you keep your work safe and that any work in electronic format has a secure and safe backup.

Be well organised. This is your chance to show that you are an independent enquirer, creative thinker, reflective learner and self-manager, and therefore will contribute towards achievement of your Personal Learning and Thinking Skills.

Be prepared with a list of relevant questions that you could ask when your teacher arranges a visit to an engineering company.

Plan ahead for your work experience and make a list of things that you need to find out about or observe, so that you make the most efficient use of your time.

Useful Links

Make good use of your work experience so as to find out as much as possible about issues that are relevant to your coursework. You may have the opportunity to work and meet with people working in design, development, workshops, test facilities, human resources and general management.

A visit to a museum or exhibition will be heplpful when investigating the impact that engineering has had on the development of the modern world.

There is a wide range of websites to help you with assessment focus 1.4, but remember to search UK sites only.

Things you might need

Your work needs to be in the form of an A4 word-processed report and should be presented as an e-portfolio. Your teacher should give you access to the required software to enable the correct presentation.

You may need access to a professional engineer who started out as an engineering apprentice.

Remember that you will have to interview people or ask questions in order to complete assessment foci 1.2, 1.3 and 1.4. Make a transcript of what was said.

Remember to maintain a focus on the work of engineers because of the impact they have on people's lives and the economic well being of the UK.

How you will be assessed		
What you must show that you know	**Guidance**	**To gain higher marks**
That the different types of engineering industries which make products and provide services can be grouped together into sectors. *Assessment Focus 1.1*	✳ Identify a number of different industry sectors and for each say what type of product they make or service they offer. ✳ Survey some of your local engineering companies, identify their operating sector and state what their business is.	✳ You need to describe two different industry sectors and give specific examples of the products they make or the service they provide. ✳ You need to explain the operating principles of some commercially available products and why they are of use to the general public.
That there are many job opportunities in engineering ranging from highly skilled crafts-persons to professional engineers who design new products and manage companies. *Assessment Focus 1.2*	✳ Visit an engineering company and talk to someone who started their career as an apprentice and has worked their way up through the business. ✳ Investigate the different types of qualifications, qualities and skills that engineers need so that they can do their jobs. ✳ Find out why professional engineers have to register with the Engineering Council (EC^{UK}).	✳ You must be able to evaluate the qualifications and skills required for a range of career opportunities in engineering. ✳ You need to describe career routes in engineering which are available at a local and a national level. ✳ You must be able to evaluate the reasons why professional engineering registration is important at both national and international level.
How engineering has improved the quality of people's lives and contributed to economic development in the UK and internationally. *Assessment Focus 1.3*	✳ Investigate examples of key achievements in engineering in the 19th, 20th and 21st Centuries. ✳ Find out about milestones in human history which have happened in the last three centuries because of the involvement of engineers. ✳ Talk to a senior design engineer and find out how technology has changed the way they work.	✳ You must be able to explain how developments in technology have impacted on people's daily lives. ✳ You should give thought to the possible downside as well as the upside of developments in technology. ✳ You need to explain how new technology has affected the local economy and economic growth of the UK.
How the rights and responsibilities of employers and their employees are safeguarded by legislation and the Department of Trade and Industry (DTI). *Assessment Focus 1.4*	✳ Interview the human resources manager of a local company and find out how they conform to the Employment Act 2002. ✳ Produce a checklist which someone starting up a business could use to make sure that they comply with workplace legislation relating to: ✳ discrimination ✳ equal opportunities ✳ health and safety ✳ family/parenting ✳ dismissal and discipline.	✳ You should be explaining why it is necessary to have employment legislation. ✳ You need to explain how the management of a local engineering company encourages its employees to work in accordance with their responsibilities.

Introduction

Design is the process of converting an idea or market need into the detailed information necessary to manufacture an engineered product or deliver an engineering service. The ultimate aim of the design process is to ensure that the client's needs are met by producing a product or delivering an engineering service of appropriate quality. The design process usually relies on interaction with a number of other people, such as production engineers, test engineers, commissioning engineers and sales personnel. As a result, a vitally important aspect of the design process is the ability to be able to communicate ideas effectively, using appropriate written and graphical skills, and in a way that can be interpreted and fully understood by other people.

In this unit you will carry out a detailed investigation of one or more engineered products. This will provide you with a valuable opportunity to understand what the product does and why it has been designed the way it has. In order to complete these tasks you will need to consider a number of aspects of the design of the product. These will include its function/purpose, the materials and components from which it is made, the processes used in its manufacture, assembly and testing, and how it is assembled, finished and finally tested. You may also need to consider the routine maintenance of the product, as well as its eventual disposal.

Before you begin a design, you need to fully understand the needs of your client or customer, and you will need to produce a detailed design specification. You will be able to experience the stages that make up the design process, and the ways in which you can develop and evaluate a range of design solutions. This is an exciting unit that will provide you with plenty of scope for trying out new ideas!

This unit is assessed by your tutor. On completing it you should:

1. Know about the construction and function of an engineered product or system.

2. Be able to prepare a product design specification.

3. Be able to prepare initial design proposals.

4. Be able to prepare and submit a final design solution.

THINKING POINTS

Think about the following key points as you work through this unit:

1. What is the design process and why is it important?

2. What factors does a designer need to consider when designing an engineered product or engineering service?

3. What are design criteria, and what should be incorporated into a design brief and design specification?

4. Why is reliability important and how is it measured?

5. Why are legislation and standards important in the design process?

6. How are ideas and solutions generated, and what techniques can be used for evaluating them?

7. What is quality and why is it important?

8. How is a final design solution presented?

Designing engineered products

Being able to design something is a fundamental engineering skill. However, it is not the end of the story. Equally important is that you should be able to communicate your design to other people: even the most basic design will be hopelessly flawed if you can't explain to people what it's about and how to make it. As an engineer, you might be involved with the design of any engineered product from an adjustable spanner to a military aircraft. This unit will help you understand the design process and how to communicate your ideas to other people.

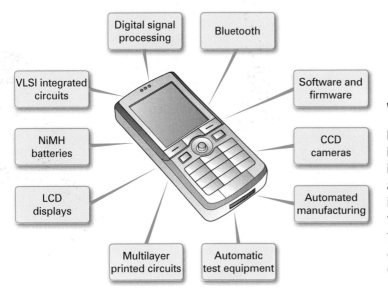

Digital signal processing · Bluetooth · VLSI integrated circuits · Software and firmware · NiMH batteries · CCD cameras · LCD displays · Automated manufacturing · Multilayer printed circuits · Automatic test equipment

What is design?

Design is the process of converting an idea or market need into the detailed information necessary to manufacture a product or deliver a service. Success involves combining skills with technologies, materials and processes, then delivering the product or service at a cost that will make it competitive with other products and services.

The design process

This means the various stages that you go through when you design something. Each stage in the process is associated with a particular phase in a design project.

Design is not always about creating a brand new product or service. In fact, in most cases it's about improving or modifying an existing product. When designing a brand new product or service, the design process will have more stages, because you usually need to consider a wider range of options and alternative solutions than when you are simply modifying or redesigning an existing product.

A typical design project involves the following tasks:

* understanding and describing the problem
* developing a design brief with the client
* carrying out research and investigation
* generating ideas
* investigating solutions and applying scientific principles
* developing an agreed set of design specifications
* communicating the design solution using appropriate engineering drawings
* realising and evaluating the design solution.

Design process – the various stages that are followed when preparing a design

Design criteria – a list of factors that need to be taken into account in a design

Compliance – ability of a product or service to meet a given standard or specification

You need to be aware that designing something involves a series of tasks and each of these forms an important stage in reaching your goal: an engineered product or the delivery of an engineered service.

Identifying a need

The design process begins when there is a need or when there is a problem to solve. The problem could be based around developing a new product or service or it could involve modifying or improving an existing product or service. You can't begin the design process unless you have a clear understanding of the problem you want to solve. So, start by asking what the problem really is—it might save you a lot of work.

What a designer needs to consider

A designer needs to consider a number of design factors including:

* functionality (what the product should do)
* aesthetics (what the product should look like)
* ergonomics (how easy it will be to use the product)
* reliability (how reliable the product will be)
* maintainability (how easy it will be to repair if the product goes wrong)
* manufacturability (how easy it will be to build and assemble the product)
* compatibility (what other products/systems it must work with)
* efficiency (how much energy the product uses and how much is wasted)
* cost (how much the product costs to manufacture, operate and maintain)
* **compliance** with health and safety and other relevant legislation.

Activity

Think about an engineered product that you use in everyday life, such as a bicycle, hi-fi system or DVD player. List three or four features or improvements that you would like to incorporate. For each, suggest what modifications or changes would have to be made. Present your work in the form of a single A4 page and include hand-drawn sketches where appropriate.

Select just one of the features or improvements that you have identified. Now think of the problem that led you to suggest that this feature needs attention. Summarise the problem in a single paragraph using no more than three sentences.

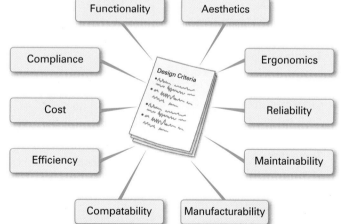

Just checking

* What typical factors need to be taken into account in a design?
* What are the stages in the design process?
* What typical factors make up a set of design criteria?

Clutch – a mechanism that causes rotation (turning) when engaged and no rotation (no turning) when not engaged

Chuck – a device that holds a drill or screwdriver bit in a rotary tool such as a drill. Chucks usually have three jaws and can be adjusted for a range of different bit sizes

Torque – a force that causes turning. The amount of torque (in newton-metres) is equivalent to the product of the force (in newtons) and the radius at which it is applied (in metres)

Transistor – a semiconductor device that can be used as an amplifier, switch or power controller

Schematic – a simplified diagram showing how various component parts are connected together and relate to one another

Product investigation

As part of this unit, you will investigate some typical engineered products. This will not only provide you with a better idea of how these products are designed and manufactured but will also help you understand some basic engineering principles.

Product investigation checklist

To help you get started with your product investigation, here is a checklist of questions to give you a better feel for what the product does and why it has been designed the way it has.

* **Function/purpose** – What is the function/purpose of the product? What need does it satisfy? How well does it satisfy that need?

* **Components** – What are the main components used in the product? Why were these components used? What alternative components could have been used?

* **Materials** – What materials are used in the product? Why were they chosen? What alternative materials could have been used?

* **Processes** – What processes are used in the manufacture of the product? Why were they used? What alternative processes could have been used?

* **Assembly/joining** – What techniques were used for assembling/joining the components? Why were these techniques used? What alternative techniques could have been used?

* **Finish** – What finish was applied to the product? Why was this finish chosen? What alternative finishes could have been used?

* **Controls/ergonomics** – What controls are fitted to the product? Why were they fitted? What alternative controls could have been used?

* **Energy source** – What energy source is used for the product? Why was this chosen? What alternatives could have been used?

* **Labels** – What labels are applied to the product? Why were they used? What information do they provide?

* **Use** – How is the product used? Are any initial adjustments necessary? If so, what are they? What safety precautions need to be taken when using the product? What user information (e.g. an instruction manual) is provided?

* **Maintenance** – What routine maintenance is required? What precautions need to be observed when dismantling the product? What components may wear out and need replacement? What service information (e.g. a service manual) is available?

* **Disposal** – How should you dispose of the product? Are any of the materials hazardous? Can any of the materials be recycled? How can you tell?

A sample product investigation

The product that you use for your product investigation should not be too complex. It should have a function/purpose with which you are familiar and it should involve a variety of different engineering materials and processes. To give you some idea of the sort of product that you should investigate (and how you should go about the investigation), we've used a low-cost variable-speed cordless drill. This has a limited number of components and it incorporates both mechanical and electrical parts, made from a variety of different materials using a number of different processes. Furthermore, because the cordless drill operates from a battery and not directly from the a.c. mains supply, it is a relatively safe product to work on.

Dismantling the variable speed cordless drill involves first removing the rechargeable battery from the base of the unit and then removing a number of self-tapping screws that retain the two halves of the plastic moulded case. Once the case has been removed, it is possible to lift out the motor, gearbox, **clutch** and chuck assembly, together with the electronic speed controller, power transistor and heat dissipater. The motor can then be separated from the gearbox, clutch and chuck assembly. To completely remove the motor, it is then necessary to release the two spade terminals that connect the d.c. supply to the motor.

Injection moulded plastic parts are used for the external case of the variable speed cordless drill and also for the case of its rechargeable battery. Plastic is tough and light and injection moulding allows very complex shapes to be produced

Electrical wiring from the electronic speed controller to the motor uses soldered tags at the speed controller and crimped spade terminals at the motor (these allow the motor to be easily detached without having to desolder any joints). The insulated copper wires are colour coded to facilitate assembly and maintenance

Internal view of the variable- speed drill showing the main components

> **Personal learning and thinking skills**
>
> The variable-speed cordless drill uses a rechargeable nickel cadmium (NiCd) battery. Use internet or other resources to obtain further information on this type of battery. Present your work as an A4 fact-sheet and include a comparison chart showing how this type of battery compares with conventional non-rechargeable alkaline batteries.

> **Activity**
>
> The variable-speed cordless drill incorporates an electronic speed controller, which uses a power transistor fitted to a heat dissipater. Why is the heat dissipater fitted and what might happen if it was not present?

Schematic diagrams

Schematic diagrams provide a useful way of describing how the various components in a product work together. They show the various component parts within a product represented by blocks, and the links between them shown by lines and arrows. You may find it useful to sketch a schematic diagram like the one shown here as part of your own product investigation.

> **Activity**
>
> The variable-speed cordless drill incorporates a **torque** adjustment feature. This adjustment varies the clutch and allows it to slip, instead of continuing to turn or twist, when a preset level of pressure is reached. Why is this feature useful and what is likely to happen if the torque is set (a) too high and (b) too low when driving a screw?

> **Just checking**
>
> ✳ What questions could be used in a checklist for your product investigation?
> ✳ What is a schematic diagram and what is it used for?

Before you can begin a design, you need to understand what your client needs. You do this by developing a **design brief** and then agreeing this with your client. However, on its own, the design brief does not include sufficient detail to chec k that each aspect of the design criteria has been met. For this, you need a much more detailed document known as a **design specification**.

The design brief

A design brief is a statement that identifies what is needed to satisfy a need or solve a design problem. The design brief must not be vague or too long, and not suggest the solution. Instead, it is for you, the designer, to suggest ways of solving the problem which fulfils the design brief and solves the client's problem in the most effective and appropriate way.

Wording is important. It's a good idea to only use simple words and agree these with your client before you begin. If the client provides you with the design brief (rather than you developing it for the client), make sure that you understand it fully before going any further.

Here is an example of a very specific design brief:

Devise a system that will raise a 500 kg load measuring 1m × 1m × 1m through a vertical height of 2 m in less than 20 s and then transport it across a workshop at a speed of 0.1 m/s over a distance of 10 m.

It's so precise that it quotes figures for the dimensions of the load, its mass, the height that it must be raised to, the distance it must travel and the speed. These figures are part of the specification for what the engineering system must do.

Design brief – a statement that identifies what is needed to satisfy a need or solve a design problem

Design specification – a set of detailed requirements that must be satisfied by the design solution

Technical specification – a set of key performance data for a product

The design specification

A specification is a set of detailed requirements that must be satisfied by the design solution. It must be written down, accurate and precise. When we use a specification to help design something, we refer to it as a design specification.

A design specification gives a very clear indication of the required performance of the product or service. It's important to establish the design specification with your client at a very early stage in any project.

The design specification is something that can be *measured*. The performance of your chosen design solution will need to be measured and compared with the specification. A successful design solution is one that fully meets or conforms with the design specification. You may well have to go back to your client if you find that your design solution only partially satisfies the design specification. Your client will have a view as to whether this can be tolerated or must be resolved.

British Standard PD6112 identifies the items that should be included in a design specification. These primary design needs are as follows.

* **Title**. The name or brief description of the product or service.

* **History and background information**. A general description of the design problem, including details of who is the intended user.

* **Design criteria/scope**. A description of each of the required features of the product or service. These may include characteristics such as size (e.g. 'must be compact', 'must be at least 2 m wide'), surface finish (e.g. 'must be smooth', 'must be water resistant'), and performance ('must be fast', 'must be powerful'). The scope may also include any special needs of the people using the product, for instance, if the users are elderly.

* **Definitions**. Definitions should be included for any terms that your client may be unfamiliar with or that may have a special meaning.

* **Conditions of use.** You will need to give an indication of how, when and in what situation the product or service will be used. For example, 'continuous use', 'light domestic applications', 'commercial and industrial use'.

* **Characteristics**. A simple breakdown of the main components, their function and any other features that the client needs to be aware of.

* **Reliability**. An estimate of the working life of the product (e.g. 4,000 hours, 10,000 operations, etc.).

* **Servicing features**. A description of the maintenance requirements and the provision for repair or replacement of materials and parts that might wear out (e.g. coolant, lubricating oil, etc) or that may need repair if the product fails or breaks down.

Other requirements may include the design factors that we met earlier:

* **Ergonomics**. Details about how users will interact with the product or service.

* **Aesthetics**. Information relating to the appearance (e.g. shape, colour, surface finish) of the product or service.

* **Safety**. Any specific safety features that relate to the product or its use.

* **Economics**. The cost of designing, producing, marketing and supplying the product or service when compared with the income generated from its sale or use.

* **Manufacture**. Information relating to how the product will be made or assembled (e.g. 'using pre-fabricated parts assembled on-site', 'supplied in knock-down form for self-assembly', etc.).

Technical specifications

You will often see technical specifications that give key performance data for a product, such as supply voltage, power, efficiency, etc. The technical specification for the variable-speed cordless drill that we met earlier is:

Voltage	12 V	No load speed	0 to 700 rpm
Battery capacity	1.2 Ah	Max. torque	9 Nm
Charge time	3 hours	Max. capacity	25 mm (wood), 10 mm (steel)
Clutch positions	5		

Reliability

Reliability is important in almost every engineering application, but in some it can be crucial. Engineering designers must be aware of when, how and why products fail.

The need for reliability

When designing engineered products, it is essential to build in an acceptable level of reliability. This can be done by applying sound design principles, using components that are 'fit for purpose' and using appropriate materials and manufacturing processes. Also, to test reliability, measurements should be carried out under conditions and circumstances that ensure the finished product performs to specification.

Some products are extremely complex, with many component parts. Failure of the product can result from failure of any component. Overall reliability will decrease unless the reliability of each component can be improved. For example, if one component in half a million would break down every hour, then a product using 100,000 of these would break down at an average interval of five hours.

Reliability means different things to different people. However, in engineering, it is usually defined as the ability of a product or component to perform its required function under stated conditions for a stated period of time, expressed as a probability.

Failure and the reasons for it

Failure is the inability of a product to carry out its specified function. It will often result from the failure of an individual component.

A product can fail in many ways, including misuse or inherent weakness in design. Sudden failures cannot be anticipated (even after a close examination of the product prior to failure). Gradual failures can be detected before they occur, by loss of performance or failure to meet some aspect(s) of the technical specification. Failures may also be partial or total referred to as (catastrophic).

Failure and lifetime

When a new purchase is made, early failures may occur due to manufacturing faults, design faults or misuse. The early failure rate may be relatively high, but fall as weak parts are replaced. There is then a period during which the failure rate is lower and fairly constant. Finally, the failure rate rises again as parts start to wear out.

Look at the 'bath tub' diagram overleaf. It has three parts:

❋ **Early failure period**. The period during which the failure rate is decreasing rapidly (sometimes called the '**burn-in** period' because the high initial failures may result from components defective in manufacture).

❋ **Constant failure rate period**. The period during which failure occurs at an approximately uniform rate. Although often shown as a straight line, in practice it will be wavy.

❋ **Wear-out failure period**. That period during which the failure rate of some items is rapidly increasing due to long-term deterioration. In good (reliable) products, it may be a long time before this period is reached.

Look at the 'bath tub' diagram overleaf.

A dip in the ocean

In some engineering applications, the consequences of a failure can be dire. For example, in the case of transatlantic submarine cable, the underwater repeaters must operate for 20 years or so without failure, because the cost of raising the cable to repair a failure could be more than £250,000 - it would be necessary to send a cable ship to the location, find the failure under several miles of ocean, supply and install a new repeater, lower the cable to the bottom again and return to port. Added to this is the loss of revenue while the cable is out of action, and the total cost might be £500,000, or more!

Activity

A marine radar system has an MTBF of 120,000 hours. Determine the failure rate expressed as % per 1,000 hours.

Activity

A low-energy lamp has a constant failure rate of 2.5% per 1,000 hours. Determine the MTTF of the lamp.

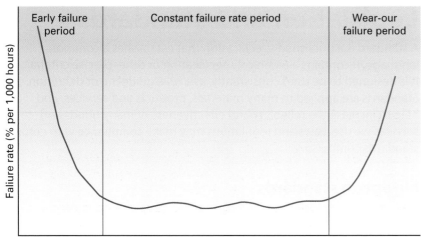

Mean time to failure

An important measure of product reliability is how long it can be expected to operate before failure. Depending on the type of product, and whether it is repairable, this is defined in one of two ways:

※ **Mean time between failures (MTBF)**. This applies to repairable items. If an item fails five times over 1,000 hours, the mean (or average) time between failures would be 1,000 divided by five or 200 hours.

※ **Mean time to failure (MTTF)**. This applies to non-repairable items and means the average time an item may be expected to function before failure. It is found by stressing a large number of the items in a specified way (e.g. by applying certain electrical conditions) for a certain period, then dividing the period by the number of failures.

When the failure rate is constant, the MTBF is equal to the inverse of failure rate. Take the case of a telephone answering machine with an MTBF of 1 million hours, used 24 hours per day:

MTBF = 1/failure rate   or   Failure rate = 1/MTBF = 1/1,000,000 hours = 0.000,001 failures per hour

This is a tiny figure. If we use 1,000 hours as a more practical time period, this would give:

Failure rate = 0.001 failures/1,000 hours = 0.1% per 1000 hours.

Since there are 8,760 hours in a year, the failure rate of our telephone answering machine (connected and used 24 hours a day) would be:

Failure rate = 0.876% per year.

Note that failure rates are often measured under ideal conditions, usually in a laboratory. In the real world, they are significantly increased by the environment and the way a product is used. These factors include temperature, humidity, vibration, etc. Temperature is usually the most critical. For example, when the surrounding temperature increases by 15°C, the reliability of a hard disk drive is reduced by about 50%. The moral here is that, if you want to keep a hard disk running, you need to keep it cool!

Functional Skills

Factory tests on a number of identical computer systems reveal the following causes of failure:

Component	Number of failures attributable
Hard disk drive	21
Memory	9
Miscellaneous	3
Motherboard	1
Modem/network card	2
Power supply	5
Processor	11

Enter this data into a spreadsheet and present it as a labelled pie chart. Which component is (a) the most reliable and (b) the least reliable? If MTBF of the power supply is quoted as 80,000 hours, what is its failure rate expressed as a % per 1,000 hours?

Burn-in – the operation of a product prior to delivery to a customer, intended to identify and reject early failures due to defective components or faulty manufacture

---

Just checking

※ Why is reliability important and how is it measured?
※ What is the difference between MTTF and MTBF?
※ What is the relationship between failure rate and MTTF?
※ Why is MTTF not the same as service life?

Keeping on the right side of the law

Working with one or two other students, think about what would happen if we were allowed to put a product on the market that nobody was sure was safe to use, that might affect the operation of other products or services, and could be disposed of safely.

A standard is a published specification that establishes a common language. It contains a technical specification or other precise criteria. It is designed to be used consistently, as a rule, guideline or definition. Standards are applied to many materials, products and services, and help to increase the reliability and effectiveness of many goods and services we use. Laws and regulations may make **compliance** with certain standards compulsory.

European standards

Equipment designed for sale or use in any European Union country must comply with the relevant **directives**: these define a set of requirements, but the standards (primarily European harmonised standards) define the technical requirements. To put it simply, a manufacturer (or the manufacturer's European representative) must first ensure that a product complies with any applicable directives before **CE marking** the product.

It is important to understand that the CE (Certification Europe) mark does not indicate conformance with a particular standard. Instead, it indicates that the product complies with all the standards that might apply to it. For example, a plastic toy and a hair-dryer will both carry the CE mark even though they meet a totally different set of standards. Now let's take a look at the CE mark and some of the directives in a little more detail.

The CE mark

Displaying the CE mark on a product (and/or its packaging) is mandatory for most types of product. The CE mark indicates that the product complies with the relevant EC directives. You will probably not be surprised to learn that there is even an EU directive that governs how the CE mark is used!

Many products need to demonstrate compliance with more than one directive. Therefore, engineering companies need to be fully aware of all of those that could potentially apply to the products they design and manufacture. Furthermore, legislation is constantly changing. A watchful eye is needed for new and revised directives.

The EMC Directive

The Electromagnetic Compatibility **(EMC) Directive** of the European Commission (EC) has widespread implications for any engineered product that uses electricity or electronics. The Directive states that products must not emit unwanted electromagnetic pollution, which might otherwise cause interference to other appliances and services. It also states that products must themselves be immune to a reasonable amount of interference.

As with all CE mark directives, the primary purpose of the EMC Directive is the creation of a single market for electrical goods throughout Europe. The protection requirements of the Directive are the means by which this is achieved. In contrast to all the other CE mark directives, the EMC Directive's primary requirement is the protection of the electromagnetic spectrum, not the safety of equipment.

Activity

Look carefully at the label for the DVD/CD rewritable drive on a computer. Who manufactured the product and where? Who designed the product? Identify the marks that show compliance with standards and directives used in (a) Europe and (b) the USA. What recycling information is displayed on the product? What special precautions should be observed when dismantling and the inspecting this product? What supply voltages and currents does the product use?

The CE mark   *The BSI Kitemark*

Low Voltage Directive (LVD)

The LVD relates to electrical equipment installation, including cables, flexible leads and wiring. It applies to equipment that operates from 50 V to 1,000 V a.c. or 75 V to 1,500 V d.c.

Machinery Safety Directive (MSD)

The MSD relates to a wide range of products that comprise a number of linked parts (at least one of which moves) and a source of energy that is other than human. Note that there is a list of exceptions to this directive, which includes tractors, military and police vehicles, as well as freight- and passenger-carrying vehicles.

British Standards

The British Standards Institution (BSI) was the first national standards making body in the world. Independent of government, BSI is a **non-profit organisation**. It is globally recognised as an independent and impartial body serving both private and public sectors. It works with manufacturing and service industries, businesses and governments to facilitate the production of British, European and International standards. BSI is also the UK's National Standards Body (NSB) and represents UK interests across all European and International Standards committees.

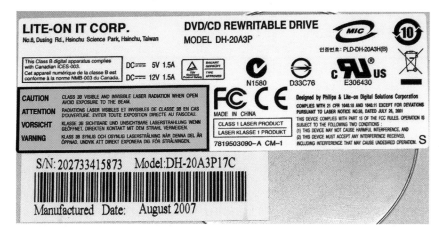

Equipment marking showing compliance with various international standards and legislation

British Standards are created by appropriately qualified and experienced people, who are brought together by BSI for the specific standard. After they agree the details, a draft of the new standard is released, and anyone with an interest is invited to comment on its content. Finally, after all comments have been reviewed, the new standard is published as a British Standard. BSI is a registered certification mark. The Kitemark and its Registered Firm symbol are instantly recognised by customers and suppliers, showing that a company is committed to improving reliability and product quality.

Compliance – being able to demonstrate that a set of prescribed criteria are met (there may be several ways of demonstrating compliance with a particular directive)

Directive – an official EU document that defines a set of requirements

CE mark – a means of identifying products that comply with the relevant European Directives (and can therefore be legally offered for sale in any EU state)

EMC directive – the EU requirement for electromagnetic emission and susceptibility

Low-Voltage Directive – the EU requirements for equipment normally operating at from 50 V to 1,000 V a.c. or 75 V to 1,500 V d.c

Non-profit organisation – an organisation in which profits are not distributed to its directors, shareholders, employees or anyone else. Instead, profits are reinvested into the services provided

Activity

Visit the BSI website at www.bsi-global.com and investigate the BSI Kitemark scheme. State the three main elements of the scheme.

Just checking

❉ What are standards and why are they important?

❉ What are EU directives and why are they used?

❉ What does the British Standards Institution (BSI) do?

❉ What is the BSI Kitemark and how is it used?

Generating ideas and evaluating solutions

An important stage in the design process is being able to generate a range of different ideas that could solve a particular problem. However, just being able to generate ideas isn't enough: you need to be able to evaluate them to find the best solution.

Generating ideas

In everyday life, ideas often seem to flow naturally. When designing an engineered product or service, this is not always the case. Furthermore, you need to spend time and effort to ensure that all practical options have been identified and explored. There are tried and tested techniques for doing this. The first is called **brainstorming**.

Brainstorming

In brainstorming a group of people sit around and fire ideas at one another. There are several basic rules:

* everyone in the group must contribute and each has an equal right to be heard

* all ideas (however unlikely or preposterous) must be treated with equal respect

* everything should be written down so that no ideas are lost (usually one member of the group is made responsible for this and ideas are recorded on a flip chart so that all can see what has been written down)

* adequate time should be set aside for the exercise and there should be no interruptions

* it is important to avoid probing ideas too deeply – this can be left until a later stage

* agree, at the end of the session, a selection (typically three or four) ideas to be considered as candidates for the next stage. These are the ideas that the group considers (by poll, if necessary) to be the most feasible in terms of satisfying the design brief. Do not, at this stage, reject the other ideas – you might need to come back to them later!

At first sight, some ideas may be considered less credible or less serious than others by some members of the group (we often describe such ideas as being 'off-the-wall'). Nobody in the group should be made to feel bad or inferior if other members of the group consider their ideas strange or unworkable. Some of the most innovative engineering projects have resulted from brainstorming sessions that have unearthed ideas that, at first sight, have been considered unworkable by the majority of those involved!

Mind mapping

Another technique, which you might have heard of, is called **mind mapping**. A mind map is a sketch that allows you to identify all the factors that need to be taken into account when developing a solution to a design brief. The name of the product or service appears at the centre of the mind map and each of the solutions and other factors are placed around it. The map can then be progressively expanded as more detail is added.

There are a number of advantages of using a mind map to generate ideas and to understand the relationship between them.

* The design brief (or design problem) appears in the centre of the map. It is, therefore, very easy to see how all the potential solutions and other factors relate to it.

* The links that exist between solutions and other factors can be immediately recognised.

* The map can be easily grasped without having to read a lot of words.

* It is easy to extend a map or add more information to it.

* A mind map can help to stimulate thought and aid understanding.

* It is often easier and faster to create a mind map than spend time putting ideas in writing!

Evaluating solutions

Having generated a number of ideas – by brainstorming, mind mapping or some other technique – you need to narrow down potential solutions to the one most likely to provide the optimum solution. The best way of doing this is to evaluate each solution against a set of design criteria, such as physical properties (e.g. size and weight), cost of materials, ease of manufacture, reliability, and so on.

When selecting the final design solution, you need to ask yourself a number of questions:

* What does the product or service do? Does the proposed solution solve the problem described in the brief?

* What materials or resources are required to manufacture the proposed solution or provide the proposed service? What tools, equipment and people are required?

* What must the product or service look like? How easily can this be achieved using the proposed solution? Are there any other aesthetic factors that need to be taken into account?

* What physical properties (size, weight, strength, etc.) are required? Can these requirements be met using the proposed solution? Does the solution meet all the physical requirements?

* Are there any problems or particular considerations that need to be borne in mind in relation to health and safety? Are there any safety or environmental issues that need to be considered?

* How easy will it be for the client, customer or end-user to use the proposed solution? Are there any ergonomic factors that need to be taken into account?

One good way of comparing a set of candidate solutions is the use of an **evaluation matrix**. This is simply a table that lists the design criteria (in rows) and each of the candidate solutions (in columns). A check mark, or numerical score, is put in each cell corresponding to your assessment of each candidate solution against each criteria.

Activity

A major retailer of auto spares and accessories has outlets in a large number of towns and cities across the UK. The company has asked you to help with the design of an auto electrical tester. They have told you that the electrical tester must:

1 be able to check lamps and fuses, continuity of wiring, and battery voltage.

2 be lightweight, compact and portable.

3 be simple and safe to use.

4 cost no more than £10 to manufacture in quantity.

Carry out a brainstorming activity with several other students to produce at least three potential solutions to the design problem. Use a flip chart to capture your ideas and then summarise them on a single sheet of A4. Finally, construct an evaluation matrix and use it to select the best solution.

An example of an evaluation matrix for the power supply of an emergency beacon that will be used in remote desert areas

Just checking

* What is brainstorming and how is it used?
* What is mind mapping and how is it used?
* What is an evaluation matrix and how is it used?

Design proposal – a detailed solution to a design problem. Usually a range of different design proposals are considered before a decision is reached on which is best

Design solution – the answer to a particular design problem, usually presented in the form of a design proposal

Candidate solutions – several potential design solutions to a particular design problem. The final design solution is selected from these

Design proposals

Development of the design brief should lead to alternative **design solutions** from which the final solution is chosen. This will usually require research and investigation to develop it, and application of scientific principles to establish the feasibility of different ideas.

Understanding the problem

Before you start on a detailed design proposal, you must devote sufficient time to clarify and understand the problem, and on collecting together information from different sources. The design brief should give a clear description of the problem, including a statement of needs and sufficient information to put the problem in context. Make sure the information in the design brief is simple, focused and concise.

You also need to identify and confirm the key features of the product or service with your client and carry out an analysis of their needs. At this stage, you should avoid stating a solution, even though this may be obvious. Keep an open mind on what is required.

The design specification should be comprehensive and should contain measurable parameters against which a prototype can be evaluated. It will normally contain information on:

* function of the product
* user requirements
* performance requirements
* material and component requirements
* quality and safety issues
* required conformance standards
* scale of production and cost.

From your technical design specification, you should normally produce at least three alternative design solutions or **candidate solutions** that offer different proposals for solutions to the product requirements. Each candidate solution should be realistic and match the requirements in the specification. You should include notes that evaluate each candidate solution's fitness for purpose.

When developing your design solutions you should consider:

* use of appropriate materials and processes
* aesthetics (i.e. what it looks like) and ergonomics (i.e. how easy it is to use)
* reliability and maintainability
* material and manufacturing costs
* compliance with relevant standards and legislation.

Research and investigation

Depending upon the design brief, your research might include:

* assessing the likely market potential for the product or service
* evaluating competing or existing products that meet a similar need
* views of potential end users.

Research might also involve investigating the range of available technologies, materials and processes. To determine the opinions of potential users, market research can be carried out through:

* questionnaires
* polls
* surveys.

Questionnaires

Questionnaires consist of a series of questions given to a particular target group. Sometimes, the target group is a representative sample of the whole population, but it could be restricted to a particular set of people (e.g. drivers over 25 with an income of over £25,000 p.a.). Every member of the target group that completes a questionnaire is known as a respondent. Questionnaires must be clear, simple to understand and should consist of mainly 'yes/no' or single word answers. If it is too long, people will not want to do it!

Polls

Polls are usually a lot simpler than questionnaires and involve respondents making a choice from a restricted selection. For example: 'Which of these four colours do you prefer?', 'Which of these three styles do you like the most?' Polls may be with individuals, or groups of people using, say, a simple show of hands. The results do not require much in the way of further analysis.

Surveys

Surveys combine some of the elements of both questionnaires and polls. They are normally carried out on a 1:1 basis. The person carrying out the survey explains each question and then notes down the response. Surveys usually take time to carry out and analyse.

Applying scientific principles

As you work towards your design solution, you will need to apply scientific principles to be sure that your idea will work as planned. These will depend on your design, but might involve determining energy requirements, power, forces, loads, speed, acceleration, and so on.

Activity

Produce three candidate design solutions for the auto electrical tester on page 41. Use a single A3 sheet for each and include sketches (showing the external appearance and internal arrangement) and write brief notes about each solution. Include references to relevant scientific principles and, where appropriate, include calculations.

Just checking

* What is a design proposal and why is it needed?
* What is a design solution and what information should it contain?
* What is a candidate solution and how is it used?
* What is the purpose of research and investigation?
* What is the purpose of market research?
* What is the difference between a questionnaire, a poll and a survey?

Quality is a key objective for most engineering companies, and design engineers need to keep this in mind at all stages of the design process. Even more important than quality is whether or not a product is safe. Designers need to build in quality and safety to ensure that their products are not only 'fit for purpose' but also won't put customers at risk or jeopardise the safety of other people.

Quality

Quality is generally defined as 'fit for purpose', which means meeting the needs of customers. Since it is the customer who assesses its value, it is really the customer who decides whether or not a company has produced a 'quality product'. Unfortunately, this can be a problem for manufacturers, because customer perceptions of quality vary: some may be more satisfied with a product than others. Therefore, manufacturers try to use more objective criteria, such as design quality and conformance quality.

Design quality

Design quality is usually the joint responsibility of a company's marketing/customer liaison and of its research and development functions. Design quality relates to the development of a specification for the product that meets a customer's identified needs.

Conformance quality

Conformance quality means producing a product that conforms to the design specification. A product that conforms is a quality product, even if the design itself is for a cheap product. That may seem contradictory, but consider the following example. A design is drawn up for a budget camera, which is made from inexpensive materials and has limited capability. If manufacture conforms to the specification, then the product is of high quality, even though the design may be of low quality compared with other more up-market cameras.

Reliability

Reliability is usually expressed in terms of service life and mean time between failure (MTBF) – a concept that you met earlier.

Service/maintenance

When (or if) a product fails, it should either be cheaply replaceable or easily maintainable. Service usually relates to after-sales service and ongoing maintenance to ensure that performance remains within specification.

Quality control

Quality control is concerned with administering all the aspects already mentioned. In the UK, there are general standards for quality systems, the

---

Making your mind up

Quality is usually something that we take for granted. In small groups, list as many reasons as you can that give you confidence that something you buy in this country is safe to use and won't harm you or anyone else! Consider at least one foodstuff, a boxed item that is sealed and a larger item like a lawnmower or car.

Consumer – the end-user of a product (a customer who purchases and then uses a product is a consumer)

Quality control – processes used within a company to ensure and maintain quality

Safe product – a product that represents minimal risk and offers consumers a high level of protection

most relevant being BS 5750 and its international counterpart, ISO 9000. The activities that make up a quality control system include:

* inspection, testing and checking of incoming parts and materials
* inspection, testing and checking of manufactured components and products
* administering any supplier quality assurance systems
* dealing with complaints and warranty failures
* building quality into the manufacturing process.

Many of these activities monitor quality after the event. Others may be carried out to prevent problems before they occur, and some to determine causes of failure that relate to design rather than manufacturing faults.

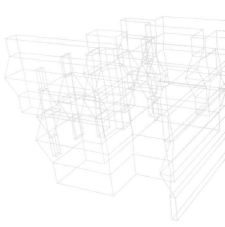

Safety

Safety is also important for design engineers to consider. As a supplier or manufacturer, you are obliged only to market safe products by:

* providing clear instructions for use, including warnings again possible misuse
* being aware of (and meeting) industry and mandatory standards
* developing product recall plans and procedures, including effective communication strategies to the public (e.g. advertisements in newspapers)
* incorporating safety into product design
* developing appropriate safety standards through product improvement
* implementing a quality assurance program that includes consumer feedback
* responding quickly and effectively to any safety concerns that arise.

If you don't comply with mandatory standards, you risk legal action. A **'safe product'** is one that, under normal or reasonably foreseeable conditions of use, presents no risk (or only the minimum risk compatible with the product's use), and provides a high level of protection for consumers. Assessment is based on such features as:

* product characteristics
* packaging and instructions for assembly and maintenance, use and disposal
* the effect on other products with which the product might be used
* labelling and other information provided for the consumer.

Activity

For each of the three candidate design solutions for the auto electrical tester (page 41) consider at least THREE aspects relating to (a) quality and (b) safety. Add these to your A3 drawing sheets.

Product recalls

When a product is unsafe or suffers from a serious defect, customers may be asked to return it to the manufacturer or supplier. Because of the negative publicity, product recalls can be extremely expensive. A list of product safety recalls is maintained by Trading Standards Central. Their website (www.tradingstandards. gov.uk) is supported and maintained by TSI, the Trading Standards Institute.

Just checking

* Why is product safety important?
* What is quality control and why is it important?
* What is a product recall and why might it be necessary?

Final design solutions

Having produced and evaluated a set of candidate solutions (and taken into account quality, safety, reliability, standards and legislation along the way), the time has now come to confirm the choice of final design solution. At this point, the design process is one of describing the solution rather than identifying it. Most engineers will be comfortable with this because it involves precise detail rather than vague ideas!

Design solutions

Your final design solution should bring together the most appropriate ideas from your candidate solutions and develop them into a workable solution that fully matches the design specification. The design solution should normally be presented in the form of a design portfolio that contains the following:

* a description of the need or design problem
* the design brief and design specification
* a description of the final design solution
* details of the candidate solutions that you considered
* details of any research/investigation that you carried out
* evaluation of the candidate solutions and justification for your final design solution
* reference to relevant scientific principles and any calculations that you made
* detailed parts or component list (including sources of any parts that will be 'bought in')
* notes, sketches, and photographs
* formal general arrangement and detail drawings (as appropriate)
* block schematic diagrams and assembly diagrams (as appropriate)
* extracts from data sheets, application notes and other documents (as appropriate)
* sketches, photographs and drawings
* any issues that relate to ergonomics, aesthetics, reliability, maintenance, standards, legislation, quality and safety (where appropriate you should include a brief section on each aspect)
* a list of references and information sources
* an **activity log**.

It is important to ensure that you provide sufficient evidence to fully satisfy the assessment requirements for the unit.

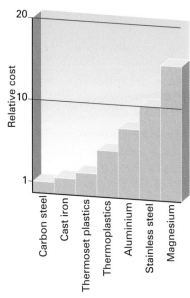

Relative costs of some materials used in engineering products

Activity log

When working on your final design solution it is important to keep an activity log, which will be used as evidence during the assessment of your work. This must record each of the stages that you have gone through to arrive at your final design solution and should include the following:

* the date on which the activity took place
* a brief description of the activity
* the outcome of the activity and any notes that you might need for future reference
* the time spent on the activity
* whether the activity was completed or not.

Production plan

Your final design solution should have sufficient detail to allow a production engineer to produce a production plan for the product, from which a number of prototypes will be produced for testing and analysis. Although you are not required to produce a prototype of your product, it would be useful to include in your design solution some suggestions for how it should be produced. Mention, in particular, the materials and processes that might be used during manufacture.

Finally, it's important to bear in mind the cost of manufacturing the product that you have designed. This means taking into account the cost of the materials that you have used as well as the cost of processing them. Wherever possible, you should ensure that costs are minimised but remain consistent with a product that is 'fit for purpose'.

Activity

Start producing your final design solution for the auto electrical tester on page 41. List all component parts and itemise the materials required. Use supplier catalogues to locate suitable parts. Indicate quantities required, stock codes and costs in your component list. Where components need to be manufactured, draw up a **production plan** that includes a list of materials and processes to be used and the sequence in which they should be manufactured and assembled. Also provide details of any quality checks or safety tests that might be needed.

Activity log – a document that summarises the work carried out in order to arrive at a final design solution

Production plan – a plan that includes sufficient detail for a product to be manufactured

Part reference	Description	Finish/notes	Quantity required	Supply quantity	Supplier	Stock code	Notes
101	9V rechargeable battery	n/a	1	1	RS	199-646	
102	Battery holder	18 gauge aluminium	1	n/a	make	n/a	See drawing 3
103	Chassis mount DC power socket	n/a	1	10	RS	486-684	
104.1-104.8	M3 pan head cross recess screw (12 mm)	Nickel plated brass	8	100	RS	483-0108	
105.1-105.8	M3 nut	Nickel plated brass	8	250	RS	483-0502	
106.1-106.4	M3 washer	Nickel plated brass	4	250	make	483-0631	
107.1-107.4	Mounting pillar (25 mm)	Threaded plastic rod	4	n/a	make	n/a	See drawing 4

Part of a parts list

Just checking

* What is a final design solution?
* What should be included in a design portfolio?

Reports and drawings

To satisfy the assessment requirements of this unit, you will need to be able to present your final design solution and explain it to other people. Your presentation may take various forms but, whatever form it takes, will probably involve a written design report accompanied by relevant drawings. Fortunately, your design portfolio will already contain most of the notes, sketches and diagrams you need to do this!

Design reports

In order to bring together key aspects of your design solution, you might consider presenting your work as a design report. This will take the form of a written document (usually word-processed) organised under a number of headings, including these.

* **Summary** – A brief overview for busy readers who need to quickly find out what the report is about.

* **Introduction** – This sets the context and background and provides a brief description of the problem you have solved. It will usually also include a statement of the design brief.

* **Main body** – A comprehensive description of the design solution including how and why it was chosen, together with details of the research and investigation that you carried out. You should also include (and comment on) the design criteria and the final design specification. Your design solution should be presented together with relevant sketches and drawings.

* **Evaluation** – A detailed evaluation of the work that you carried out, including any problems that you encountered and how you solved them. You also need to be able to justify your final design solution in relation to the range of candidate solutions that you considered.

* **Recommendations** – This section should provide information on how the design solution should be implemented (including, for example, information on the materials and processes used for its manufacture). You may wish to include details of any tests or measurements that should be made to confirm the operation of your product. You can also include any modifications or changes that you recommend or any additional functions or applications that your product might have.

* **Conclusions** – You should end your report with a few concluding remarks about your design solution and how effective you think it's likely to be at satisfying the original need or solving the original problem.

* **References** – This section should provide readers with a list of sources for further information relating to the scientific principles or technology used, including relevant standards and legislation.

Finally, it is important to take care how you present your work as a written report.

To avoid confusion, the normal conventions of grammar and punctuation must be used. Words must be spelled correctly, so do make use of a dictionary or, if you are using a word-processing package, use the spellchecker. Never use jargon terms and acronyms unless you are sure they will be known to readers.

Layout is important, so use numbered sections and paragraphs and try to keep sentences short and to the point!

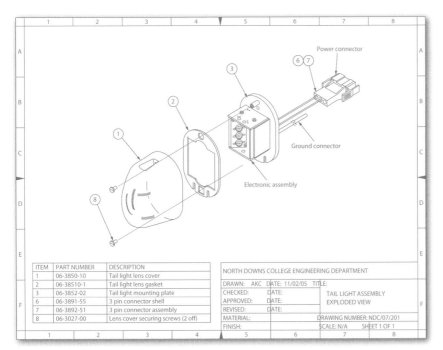

ITEM	PART NUMBER	DESCRIPTION
1	06-3850-10	Tail light lens cover
2	06-38510-1	Tail light lens gasket
3	06-3852-02	Tail light mounting plate
6	06-3891-55	3 pin connector shell
7	06-3892-51	3 pin connector assembly
8	06-3027-00	Lens cover securing screws (2 off)

NORTH DOWNS COLLEGE ENGINEERING DEPARTMENT

DRAWN:	AKC	DATE: 11/02/05	TITLE:	
CHECKED:		DATE:		TAIL LIGHT ASSEMBLY
APPROVED:		DATE:		EXPLODED VIEW
REVISED:		DATE:		
MATERIAL:			DRAWING NUMBER: NDC/07/201	
FINISH:			SCALE: N/A	SHEET 1 OF 1

A typical general arrangement diagram for an electronic tail light assembly for a light aircraft

Drawings

Your final design solution (as well as your written report) will need to incorporate a number of drawings of different types. These might include:

* **Schematic diagrams** – These use standard symbols to show how things are connected together. There are several types of schematic diagram, including those used for electrical and electronic circuits, pneumatic (compressed air), and hydraulic (compressed fluid) circuits. Schematic diagrams are often used to show how the functional blocks relate to one another. They are particularly useful for showing how signals or information 'flow' from one stage to the next.

* **General arrangement drawings** – A GA drawing shows all parts used in an assembly. They are often listed separately in a table, together with the quantities required. The parts are usually 'bought-in' as 'off-the-shelf' items from other suppliers, and supplier catalogue references may be included. Separate detail drawings will be required of components to be manufactured as special items. Dimensions are not usually given, although overall dimensions may be given for reference, if the drawing is of a large assembly drawn to a reduced scale.

* **Detail drawings** – Detail drawings provide all the details required to make an individual part or component, including shape, dimensions, materials, treatments and finishes.

* **Exploded views** – These are pictorial representations of a product showing how the parts fit together. Each part or component is shown separately but in approximately the same physical relationship as when assembled. This gives an idea of how something is put together or dismantled. A service or maintenance engineer has only to take a look at an exploded diagram to see how the various parts fit together.

Activity

Produce a full set of drawings (including a schematic diagram, general arrangement diagram, and detail drawings) for the auto electrical tester on page 41.

Just checking

* What is a design report?
* What is a schematic diagram?
* What is a general arrangement drawing?
* What is a detail drawing?
* What is an exploded view?

Presentation techniques

The ultimate challenge for a design engineer is to persuade other people that your ideas are workable. At the completion of this unit, you will be required to present your final design solution to a critical audience. For most people, this can be quite a challenge but it provides an excellent opportunity to practice your presentation skills, and to learn how to respond to questions and suggestions made by other people. This is yet another skill that designers need to have!

Delivering your oral presentation

Your oral presentation will normally last between 10 and 20 minutes. To make your talk more interesting, it should be supported by one or more of the following visual aids:

* **flip chart**
* **Microsoft PowerPoint presentation**
* set of **overhead projector** transparencies
* models or product samples.

In addition, you should make sure that your design portfolio (including the activity log, design report and set of formal drawings) are available, in case you need to refer to them during your presentation.

The art of good delivery

Your presentation should be delivered in a way that is appropriate to your audience. It should be brief and to the point but must cover all the main points that make up your design solution. It must be interesting and appropriately paced so that the attention of the audience does not wander.

When delivering a verbal presentation, it's also important to be a good listener, to be able to respond to questions or queries raised by your audience. At the outset, you might wish to indicate whether you will take questions during your talk or at the end. There can be good reasons for either but, in the end, how you deal with questions is up to you!

Flip chart – a set of A2 or A3 sheets that can be flipped over to reveal their contents. Flip charts are quick and simple to use and are ideal for use with smaller audiences

Microsoft PowerPoint presentation – a presentation using a software package that forms part of the Microsoft Office suite. It can be delivered to a large audience using a screen and projector connected to a computer

Overhead projector – a stand-alone projector that uses A4 transparencies or rolls of acetate film. They can be used with large or small audiences, but can sometimes be crude and hard to read

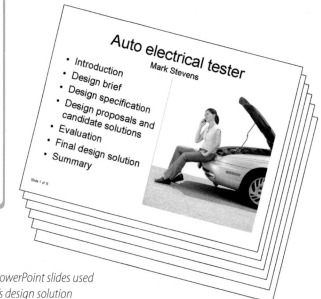

An example of a set of PowerPoint slides used for presenting a student's design solution

Use what you've got!

Everything that you need to include in your oral presentation should already exist in your design portfolio. All you need to do is summarise it and present it in a way that your audience can quickly and easily grasp. In most cases, you should supply your audience with a set of notes or printed handouts. These can be based on your PowerPoint slides or hard copy print-outs of overhead projector transparencies, augmented by sketches and presentation drawings as appropriate.

A typical oral presentation runs along the following lines (you might like to consider one flip chart page or one PowerPoint slide for each).

❋ Introduce yourself and say who you are.

❋ Outline the need that your product satisfies and describe the original design.

❋ Explain the design specification.

❋ Explain the design criteria, design proposals and candidate solutions.

❋ Explain the method(s) used to evaluate the design proposals and how you justified the final design solution.

❋ Explain the final design solution.

❋ End with a brief summary of your work.

Finally, do not forget to thank your audience for listening and invite any further questions.

Practise, practise, practise

It's important to rehearse your presentation a few times before the 'real thing'. When you do this, you might want to enlist the services of a few friends or family members to act as a 'tame' audience. They will give you some pointers on how your presentation can be improved.

Good luck with your presentation!

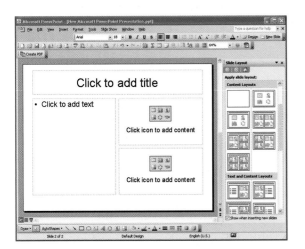

PowerPoint provides you with a wide variety of templates that will help you lay out your slides

Activity

Prepare and deliver a PowerPoint presentation for your final design solution for the auto electrical tester on page 41.

Just checking

❋ What visual aids could be used to present a design solution?

❋ What should be included in a design presentation?

Unit 2 Assessment Guide

In this unit, you will investigate how products are designed so that, when they are bought by a customer and put into service, they function as expected and to specification. Starting with a customer design brief, you will produce a product design specification (PDS) and then a number of alternative design proposals which meet the requirements of the PDS. You will then investigate the techniques used to present and compare design ideas so that a decision can be made on which one to develop. Your teacher will play the part of the customer.

Time Management

Manage your time well as this unit has a number of different components that will have to be researched. Ensure that you keep your work safe and that any work in electronic format has a secure and safe backup.

Be well organised. This is your chance to show that you are an independent enquirer, creative thinker, reflective learner, self manager and effective participator and therefore will contribute towards achievement of your Personal Learning and Thinking Skills.

Be prepared with a list of relevant questions that you could ask if your teacher arranges a visit to an engineering company

Plan ahead for your work experience and make a checklist of things that you need to find out about or observe so that you make the most efficient use of your time.

Useful Links

Make good use of your work experience to find out as much as possible about issues that are relevant to your coursework. If you have the opportunity to meet with people who work in design and technical sales, ask them about their relationship with customers and how a job is progressed through the factory. Make notes about what you find out.

As there are many websites giving information about design standards and legislation, just concentrate on using UK data.

Things you might need

Your work needs to be in the form of a process portfolio which contains sketches, diagrams, formal drawings and a written report. Drawings can be produced by hand or using a commercial CAD system.

Your teacher should give you access to drawing equipment, a CAD system and presentation software, such as PowerPoint.

Access to a professional working in the design department of an engineering company would be a good way of finding out more about how the design process operates.

Access to an overview of the standards and legislation which apply to the design of the product you are investigating.

Remember to maintain a focus on design as a process– the starting point is a customer design brief and the finish a final design solution.

How you will be assessed

What you must show that you know	Guidance	To gain higher marks
That by dismantling and reassembling a product or service you can find out how it is constructed and operates. *Assessment Focus 2.1*	✳ Dismantle and reassemble a product or system and make notes and sketches which describe how it is constructed. ✳ Obtain witness statements and take photographs to prove that you carried out the practical work. ✳ Investigate the function/purpose of the product or system.	✳ You need to investigate how the product or system operates. ✳ You must be able to evaluate the product's fitness for purpose and whether it performs to specification.
How to use a customer design brief (CDB) as the starting point when designing a new product or system. *Assessment Focus 2.2.1*	✳ Meet with your customer and find out what they want you to design. ✳ Working from the customer design brief, you should identify some physical features about the product/system such as its mass, dimensions, colour and surface appearance.	✳ You must identify a range of physical features which relate to the requirements of the CDB. ✳ You must identify features which relate to the performance of the product/system e.g. operating voltage, speed of rotation. ✳ You must identify features relating to reliability e.g. service life and what happens if a fault occurs.
How to develop a customer design brief into a detailed product design specification (PDS). *Assessment Focus 2.2.2*	✳ Using the CDB and physical/performance data which you have already put together, prepare a PDS for the product/system which you are investigating. ✳ Discuss what you have produced with the customer.	✳ You must provide evidence that you have investigated the standards (e.g. BS) and legislation (e.g. European Conformity) which relate to the product/system you are designing. ✳ You must give details about manufacturing costs and equipment needed to make the product/system.
How to prepare design proposals which meet the requirements of a PDS and can be compared against each other. *Assessment Focus 2.3*	✳ Produce three proposals using free hand sketches, circuit diagrams, flow charts, etc. ✳ Add written notes and calculations to your design proposals. ✳ Present your proposals in a format which allows them to be compared. ✳ Choose one proposal for development.	✳ You must provide evidence that you have used a recognised evaluation technique when choosing which design to develop further. ✳ You must fully justify your choice of which proposal to develop
How to produce a final design solution and submit it to the customer. *Assessment Focus 2.4*	✳ Produce detailed sketches, drawings, circuit diagrams, flow charts, etc. by hand or using software packages. ✳ Put together supporting documentation such as notes, performance calculations and costs.	✳ Make an oral presentation to an audience explaining your design and how it meets the requirements of the customer design brief and PDS. ✳ Include in your design portfolio a log book which contains notes about things you did to progress your work e.g. discussions with the customer, research, timeline, etc.

Introduction

The use of computers is of vital importance to the way we live our lives today. In the modern global economy engineering industries have had to embrace computer technology in their manufacturing systems, products and services in order to remain at the forefront of innovation and design. Continuing advances in computer-aided systems can give those that invest in their development a competitive edge, enabling them to achieve greater productivity, reliability and quality. We as consumers demand this delivery, reliability and quality all at as low cost as possible and on time.

In this unit you will discover how computer systems are applied to a range of aspects of engineering manufacturing from product design and development to automated packaging, to maintenance operations. You will also learn how many everyday products use microprocessor technology. In this topic you will look in detail at a range of microprocessor control systems found in our everyday life: things like microwaves, motor vehicle management systems and mobile phones. As well as motor vehicle computer diagnostic testing you will look at other developments of 'plug-in' diagnostic systems, such as hand-held terminals that aid maintenance engineers.

How you will be assessed

This unit will be assessed by your tutor who will set an assignment for you to complete. It will focus on giving you an overall understanding of the way computers contribute to the world of engineering. As such, you will be assessed through an assignment that will give you opportunities to demonstrate what you know about computer applications in process control and manufacturing, that you can use computer-based systems to solve engineering problems, what you know about computer aided technology in maintenance operations and computer communications systems.

After completing this unit you should be able to achieve the following outcomes:

1. Describe and review how two different industries use computers in process control and manufacturing.

2. Set up and review the use of appropriate computer-based equipment to solve a given engineering problem.

3. Describe and review how microprocessor systems and their component parts control the features or actions of at least two consumer products.

4. Describe how two different computer applications can be used in a given maintenance operation and show how they can be used in the diagnosis of faults.

THINKING POINTS

Obviously, computers were used in manufacturing and engineering a long time before they became common place in our homes. However, if you know someone who was born before 1966 and ask them if they used computers at work when they first left school, they will probably say no. Since those days the rise and introduction of the use of computers to do all sorts of things has been tremendous. We all know about the use of computers to give us access to the internet, but where are computers used to help engineers? You would probably come up with design, but think about the many other ways. You will learn about these other applications in this unit. Is the use of computers always good?

The delivery man who comes to our door and asks us to sign to prove we have received the parcel asks us to do so on a hand-held screen: within minutes the sender would know that their parcel has been delivered. Where else in engineering would a hand-held computer be used? How are applications of engineering controlled? How are the processes controlled? Do you know any motor vehicle mechanics? If you do, how do you think their job has changed with the widening use of computers and their technologies?

Safety and fast manufacture are both important in engineering: ask yourself, how have computers helped? Look at the appliances in your kitchen at home, what controls them, apart from you mum or dad?

Finally, as you work through this unit, think: would you like to live in a world without computers?

Integration of computer technology

Computers have helped to change and develop engineering and many of its applications. In this topic, you will learn something about the first computers. You will also look at how they shape our life today.

A short history of computers

A computer is a programmable electronic device that can store, retrieve and process data.

Initial use of computers can be traced to Colossus and the Harvard Mk1. The Colossus was built in Britain at the end of 1943, designed to crack the German code system on radio messages that were being intercepted. Harvard Mk1 was built in the USA with the backing of IBM in 1944 as an all purpose electro-mechanical programmable computer.

Early computers were just large calculators, able to do lots of repetitive sums very quickly but difficult to set up for each task. Instructions had to be written in 'machine code' that detailed each instruction for the computer central processing unit (CPU) to carry out. They were specific to each computer design. 'Assembly language' speeded this up by using simple codes that were then turned by the computer into specific, often repeated actions. Programming was less specific to individual computer hardware design but immediately created an overhead in the computer, which now had to be able to interpret all the assembly code instructions and turn them into machine code for the CPU.

Higher level languages started to be developed in the early 1950s. These allowed 'source code' to be written in a much more human readable form, which could then be turned by the computer into machine code. The languages were specific to particular requirements and were still very time consuming to use, by software engineers who specialised in them. Examples are FORmula TRANslator (FORTRAN) for scientific use and COmmon Business Oriented Language (COBOL) for business use. Data was often created external to the computer as magnetic tapes or punched cards (more on this below). Hence, early commercial use was mainly in banking, or for company payrolls.

In the 1960s, operating systems were introduced, which interfaced between the user and the CPU to carry out repetitive tasks like file management, processing data and handling user inputs (e.g. from a keyboard). Programming languages became more versatile, closer to human understanding and with agreed international standards that made them less dependent on specific computer platforms; but they remained time-consuming to use and many developments overran or failed due to software problems. Operating systems improved in parallel.

In the 1980s, there were major efforts to improve on the productivity and reliability of software development, with formal methods, structured

Input – whatever goes into a computer, from commands entered from the keyboard to data from another computer device

Output – anything that comes out of a computer. Output can be meaningful information or encrypted code and appear as binary numbers, characters, pictures, printed pages, etc.

programming and object-oriented programming. However, all of these advances increased the internal overheads in a computer. If you buy a new desktop computer today, before you have added a single programme of your own or any data, it will already be using several hundred times the memory that was even available on such a computer 20 years ago.

Computers in engineering

Computers only became cost-effective as an aid to engineering in the 1970s but their use has expanded rapidly. Some of the ways they now assist are shown in this table:

Information flow	Ability to send information to others instantly, and receive a response almost immediately
Problem solving	Software packages designed to simulate situations, such as stress analysis
Storing and sorting data	Databases, mailing lists, inventory, client/supplier lists, etc.
Aiding manufacture	Control systems designed to undertake set tasks following an input, resulting in an output
Innovation and design	Numerous 2-dimensional and 3-dimensional drawing packages, enabling information to be easily edited and transferred
Automation	Controlling systems to enable complex tasks to be carried out
Communication	Video conferencing, being able to speak to customers, employees, etc. in real time

Apart from the stress in the materials and its joints, what else on a plane do you think might need to be analysed before actual flight takes place?

Capturing, storing and outputting computer data

Progress in using computers has developed in parallel with methods of entering, storing and outputting data. Early computers required all data to be entered as computer code through a keyboard. This progressed through paper tape and punched cards for faster input and output of repetitive information. Improvements came with the use of magnetic tape and then floppy disks, but little could be stored within the computer itself. The keyboard is still used to enter data, but for relatively limited purposes, such as word processing or control of other activities.

Even a stand alone PC can now accept data in many different ways, storing large quantities for rapid access within itself and outputting it to a screen, printer, loudspeaker or other storage devices. Several computers can be networked and data shared between them.

Stress analysis

Process control

One of the technologies used to reduce costs and increase productivity within engineering is process control. This covers any regulated task or operation – even something simple like cleaning your teeth. How you squeeze the toothpaste on to a toothbrush controls the amount and spread of the toothpaste, depending on how hard you squeeze the tube and how quickly you pass it along the bristles. More complex processes benefit from computer support.

Principles of process control

Process control can be complex, requiring numerous tasks or operations to be controlled at the same time. We can make each task simpler to understand by deciding what the action is intended to achieve. Then we ask 'if' something happens, 'then' what would need to be done to control it. We call this **negative feedback**.

Think of a central heating thermostat. This measures the temperature and compares it with the temperature you have set. If the temperature is too high, it switches off the boiler. If it is too low, it switches the boiler back on again.

How accurately we maintain our desired value depends on how sensitive our measurement of the error is and how quickly we can reverse it. Thus, the process oscillates above and below the desired value within accepted **tolerances**.

Application within industry

Computers are used widely in manufacturing to help control a wide range of processes. These include:

* controlling the flow of items on a production line
* controlling the temperature and flow of chemicals in a chemical plant or food production
* monitoring work in progress at all stages in almost any manufacturing process.

Monitoring progress enables the manufacturer to know what is happening at all times, to identify problems as soon as they occur (or even identify trends that show they are about to happen) and to keep customers informed on the progress of orders. It enables automated maintenance of stock levels for 'just in time' production and has a major input to maintaining safety and efficient use of machinery.

You will explore some of these areas in more detail in other topics. Here you will look at how the food industry might use process control. An example, similar to spreading toothpaste, is the process of applying identically patterned ribbed chocolate to sweet bars.

The image shows just one of the production stages of producing a chocolate sweet – applying a consistent thickness and pattern of chocolate. Customers expect each one to be identical. This indicates high quality and that they are getting what they paid for.

Where do we use process control in our daily lives?

Think about a simple function, such as preparing a meal. Break this down into separate actions that make up the process. What is each action aiming to achieve? Consider what would happen if any processes were incorrectly completed.

Functional skills

Using the steps for preparing a meal, add the control aspect for each task or operation by asking 'what happens if …?' and deciding what would be needed to control it. Once complete, write this up and give it to your teacher. As an additional task, add what you would expect to happen as the outcome at each stage.

Applying identical patterns to sweet bars

Manufacturers must consider the cost of achieving a product consistent in appearance, weight and taste against the possible loss of sales, if these are not maintained.

There are many steps in the full manufacturing process for example:

❋ controlling the quantity and quality of the ingredients

❋ how they are combined and treated to make the product

❋ wrapping each sweet individually

❋ bulk packaging for transportation to sales outlets.

Chocolate being mixed under tight temperature control

Activity

You are an engineer who has been asked to report on what temperature controls will be required in a new chocolate factory. You first have to understand how chocolate is manufactured and how critical temperature control is at the different stages.

If you look up 'chocolate making' on the web you will find descriptions of how chocolate is made (for example: www.fieldmuseum.org/chocolate. Think of the processes requiring heat (or lack of it) and then estimate how critical the temperature must be to each. This could start with refrigeration of ingredients, then temperature control during different manufacturing stages and, finally, making sure that the product remains in good condition during storage and transport. For each process, ask 'what if the temperature becomes too high or too low?' and 'what could happen then?'. Write up your findings in a report of not more than 200 words.

As an extension, think where computers could help at the different process stages.

Negative feedback – measurement of a change in a process, comparing it with the desired value and using this to reverse the change back towards the desired value

Tolerance – acceptable variations from a desired value

Just checking

❋ Why does process control using negative feedback oscillate around the ideal value?

❋ Where could you find process control being used within engineering and manufacturing?

❋ Why could temperature control be critical in a manufacturing food process?

Can robots run our homes?

We are used to labour-saving devices, like vacuum cleaners or washing machines. Computers can now be used to control heating, lighting, even drawing curtains. Can you think of any other home chore that could be completed by robots? What if something went wrong?

Computer numerical control (CNC) – a machine that can be reprogrammed and controlled by computer to complete a sequence of operations

Robot – a machine that functions in place of a person or persons

Payback time – the time it takes to cover the cost of new equipment and start to make a profit from it

An industrial robot in use

Automation and computers

Automation preceded the arrival of computers. However, since the 1970s, computer technology has helped companies and whole industries to revolutionise the way they work. Robotics, in particular, has reduced reliance on human skills and enabled large increases in the output from factories. Provided they are properly maintained, robots are happy to work 24 hours a day, producing identical items and never getting bored or careless!

Automation in industry

Industry has been quick to adopt new technology as a way of speeding up manufacturing and reducing costs, particularly the cost of employees. Assembly lines in the 1930s linked individual machines, each completing a single task and probably operated by one person. Individual machines were automated from 1940 onwards with 'numerical control', able to complete a sequence of preset functions. **Computer numerical control** (CNC) machines made it easier to change and control these functions, and many are still in use.

Industrial **robots** are essentially computers controlling multiple tools. They can be programmed to carry out a complex sequence of operations. Computers may also monitor and test for quality, and control the flow in of components and the flow out of completed products. Some robots are built for specific tasks, while others are effectively bought off-the-shelf and programmed to each factory's requirements. They can also be reprogrammed, and tool sets changed, to make new products. Factories could, in theory, operate by themselves, controlled remotely from computer workstations, except for visits by maintenance staff and those transporting goods in and out.

Industrial robots

The frequent image of a robot in newspapers and books is of something that looks and thinks like a human. Robots that move around and can carry out tasks that are done by humans are being developed, particularly to operate in difficult environments, including space. However, the term comes from the Czech word 'robota', which means 'dull, repetitive activity'. The majority of robots are used in industry in fixed locations, having an arm that can be moved to a required location and carrying tools at the end to complete a job. They can be defined as 'reprogrammable, multifunctional manipulators designed to perform automated tasks'.

To meet the international standard for classification as an industrial robot, they must be 'programmable in three or more axes'. We can define the location of any point in space from another one by distances along three axes at right angles, usually called X, Y and Z. (Think about standing on a road and wanting to reach the top of a platform off to one side. You can walk along the road until level with the platform; turn at right angles and walk until you are directly underneath it, then turn another right angle to climb up to the platform.) An object, such as the tool on the end of a robot arm, can then rotate around three other axes. In a plane or boat these are called roll, pitch and yaw. One way robots are defined is by the number of axes that the robot arm and the tool at the end of its arm are free to move – called 'degrees of freedom'.

There are then other capabilities that define how versatile a robot is and the tasks it can perform. These include what tools it offers at the end of its arm, its working envelope (the space in which the tool can move), how the robot arm and tools are manipulated and driven, speed of operation, accuracy, etc. They are grouped into the following, each with advantages and disadvantages:

Then and now

* Cartesian robot
* Parallel robot
* Articulated robot
* SCARA robot (selective compliance assembly robot arm).

Disadvantages of automation

Automation and robotics enable manufacturing and engineering to be achieved quicker, cheaper, with more reliability and with greater flexibility. However, there are disadvantages:

* Initial set up costs. Robotic machines are expensive to purchase, set up and install. The infrastructure to feed in components, etc. may be complex.

* Maintaining competitive advantage. Competitors may buy newer, more efficient machines. Each factory owner has to keep asking 'when do we replace ours and how to do it without disrupting manufacture'?

* The total workforce may be reduced, creating social problems.

* Those required to operate the machines may need higher skills and specialist training.

* Robots are computer-controlled, so security must include the possibility of electronic attack, especially if using remote computer control.

Personal learning and thinking skills

Imagine what a vehicle factory would have looked like before computers and robotics. How many people would be employed? How big would the site be? How many vehicles would be produced?

Activity

Research the four categories of industrial robots. Identify at least one leading product in each. List the main advantages and disadvantages specific to each category.

Initial cost for robots is high. As an extension, see if you can find any information on '**payback**' time?

Activity

Create a list of the advantages and disadvantages if we were only to use computer technology and robotics to manufacture products. Agree this list in class. Then allocate each advantage a score from 1 to 10, and each disadvantage from -1 to -10, for how serious you think each is. Add them together and compare your totals with others in the class. Has this changed your opinion on using computer technology within engineering and manufacturing?

As an extension, you could produce your list as a spreadsheet with the scores in a column, which automatically adds the scores as you enter or change them.

Just checking

* When did computer technology start to affect automation?
* What security problems could there be with automation?
* Do robots increase or decrease jobs in industry?

Computer-aided design and computer-aided manufacture

Even a simple object may require several engineering drawings as 2- and 3-dimensional images (see Unit 4) before deciding on a design and providing information for manufacture. A car or aeroplane would take thousands of drawings and mock-ups before any start was made on building a prototype. CAD software allows images to be drawn directly on a computer screen and refined until they contain all information necessary for manufacture. CAD/CAM software takes this a step further, right through into helping to control manufacture.

Computer aided design (CAD)

CAD replaces the time-consuming effort of producing innumerable engineering 2- or 3-dimensional drawings before deciding on a final design. When first introduced, it just speeded up the drawing process, enabling lots of drawings to be produced, refined and scrapped before printing out final versions for manufacture.

With modern computer graphics, even complex items can be drawn as separate components, assembled into larger components, viewed in 2- or 3-dimensions as a solid figure or **wire frame**. Images can be rotated on the screen, the computer automatically redrawing the image with all dimensions correct from the new angle. It can also be sliced through and viewed from the inside.

CAD can effectively produce a prototype of a new design in software to be viewed on the computer screen. When used with advanced simulation packages, the design can even undergo extensive testing before anything is actually built. Three-dimensional models can be generated directly from the screen by sending data to a machine which lays down layers of material (for example, plastics or wax) to build up a complex shape. The process is called Rapid Prototyping, but it can take several hours to complete a compicated component. However, this is much quicker than setting up machinery to make the component in metal.

Computer aided manufacture (CAM)

In the previous topic we mentioned 'numerically controlled' machines, able to carry out a set of repetitive actions without human intervention, which were then followed by CNC (computer numerically controlled) machines. These are more flexible because the sequence of operations can be changed. The machine reads instructions that have been programmed in a fairly simple computer language, such as **G-code**. Typically, the instructions drive the machine's tools in a given direction, for a specified distance and at a set speed.

CAM is used when the control goes beyond just generating instructions for a CNC machine, to where the computer is being used to control a part of the whole manufacturing process.

CNC router with multiple router bits

Computer-aided design and manufacture

Computer control of machines has been progressively improved. The term CAD/CAM is used to cover a wide range of software packages that go beyond programming individual machines, or control of small elements of manufacture. In theory, it is now possible to use an advanced CAD/CAM package to design a product and to translate this directly into a series of processes to manufacture the complete product. Many also claim to be simulation packages, enabling the product to be viewed on the screen and extensive testing to be carried out.

Wire frame image – drawing in a CAD or simulation package showing the outline of the image, so that it can be looked through to show internal elements

G-code – a computer language for writing code to control CNC machines

Computer simulation – a computer model of an object or process

Setting up a computer-controlled milling machine

Activity

As a group no bigger than three, produce a presentation on G-code. Some areas to investigate are:

* Who first developed G-code?
* What is the commonly agreed standard for a CNC language (G-code)?

As with all languages, there are variants. Why have these happened?

Activity

Search for CAD/CAM on the web and identify ten leading software packages. Make notes on what they claim to do. Select five of these and produce a spreadsheet with these as column headers. Down the side, show the principle uses that you have found for CAD/CAM packages, such as 3-D visualisation based on 2-D drawings. Then on the spreadsheet indicate whether your chosen package performs this function, allocating them either a 0 or 1. Add these up to show which package is most comprehensive.

As an extension, see if you can find any independent reports on how easy each package is to use. This may indicate additional functions. You could now grade each package from 0 to 10 for each function you have found. The totals would give a much better indication of their relative capabilities.

Just checking

* What is the difference between a CAD and CAD/CAM package?
* How are computers used to control CNC machines?
* Why do we find different versions of computer programmes, such as G-code?

Detection – identifying that something is present or absent

Recognition – identifying the type of object present

Quality control – activities or techniques to check a product or service is fit for purpose

Reject – item of inferior quality discarded after checks

Automated systems for detection and recognition

Most of us take for granted that we can look at something from different angles, distances and light levels and still recognise what it is. We even recognise faces in a crowd, though we still get it wrong! We call this 'pattern recognition' and it is actually a very complex task, especially for a computer. Earliest use on the production line was just **detecting** whether or not something was there. However, industry has rapidly advanced in using automation to not just detect but also recognise objects, as sensor and computer technology has improved.

Manual recognition

If we consider a manual assembly line, each person on this actually has several different tasks to complete. For example, each component may need to be detected, recognised, checked for quality, married correctly to other components and the finished item either passed on to the next process or discarded.

We can draw this as a flowchart. The example shows part of one, after initial checks of components have been completed.

Quality control is very simple here. Does the new piece fit correctly? If so, then the operation is completed. If not, the component is discarded and a new one selected. Most processes are more complex.

Automated detection and recognition

Automated detection systems are now commonplace, from burglar alarms to infrared sensors that automatically open doors when someone approaches. Automated recognition is also being introduced in many areas, like number plate recognition on motorways, or scanning luggage for weapons. Number plates are relatively easy, because a computer only needs to recognise letters and numbers as patterns and compare them with stored images until it finds a match.

Recognising the type of car, or weapon, is much more difficult without human assistance. The computer receives a 2-dimensional picture, including the background. It must first isolate the vehicle or weapon and then compare the image with a database containing images of every possible vehicle or weapon shown from different angles. The task is easier if we can combine the image with other information. For example, a metal detector might help to indicate that the image was a weapon and not a plastic toy.

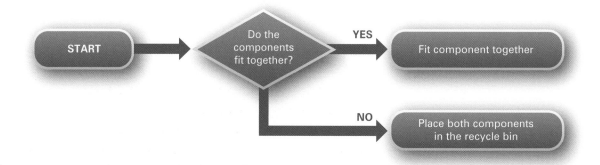

Production lines

Recognition systems are used increasingly during manufacture: for example, for sorting objects and checking quality.

For quality control, the computer can be programmed to take different actions depending on results. For instance, one occasional component out of tolerance may become a **reject**, but four such components in 100 could stop the assembly line.

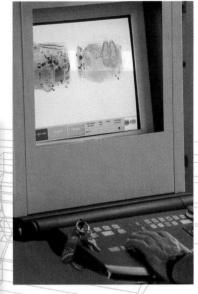

Activity

Screws are mass-produced, sorted and boxed automatically. Customers are not happy if some are the wrong length, have uneven pitch or no slot for the screwdriver!

Look at several different screw types. What characteristics must be checked? List these and then consider how this could be done using automated systems depending on automatic recognition. You will find that it is not as easy as it seems. For example, think how a computer could measure length, if working from a camera image that is recording several screws at once, all lying at different angles as they pass under the camera lens.

Produce your results as either a table or spreadsheet. List the characteristics to be measured, what automated systems could be used, the problems in doing this and any suggestions for overcoming these.

Detection system for luggage.

Activity

In small groups, investigate an industrial detection and recognition system. Decide what is being detected, how accurately it must be recognised, and what happens to the product, component or service once the system finds a problem.

Produce this as a presentation. Each member of the group should deliver part of the presentation.

As an extension, use PowerPoint to produce two automated presentations: the first showing the system when no faults are being detected; the second showing what happens after finding a fault.

Various screw types

Just checking

* What is the difference between detection and recognition?
* Why do computers find recognition difficult?
* How does industry use such systems during production?

Intel Pentium 4 microprocessor

An introduction to microprocessors

Think of thousands of processes all being produced by something no bigger than a thumbnail – science fiction becomes science fact. The Harvard Mk1 computer was enormous, taking up several large rooms, but had less processing power than today's digital watches. Microprocessors are tiny, but are the brains behind super computers and the driving force of technology today.

So what is a microprocessor and how does it work?

A microprocessor is a **chip** (an integrated circuit formed on a single slice of silicon) that contains all the arithmetic, logic, and control circuitry to perform all of the functions associated with a **CPU** (central processing unit), which is at the heart of every computer. The very first microprocessor was the Intel 4004, introduced by Intel in 1971.

A chip is a very thin slice of silicon a few millimetres across, chemically etched with **transistors** and other components. The transistors provide the computing power. The Pentium 4 chip has 42 million transistors. These form multiple circuits that are led out to connectors. Designers are provided with information so that they can connect the chip up to meet their specific computing requirements. For most tasks, only a tiny fraction of the components on the chip are used, but this is still much cheaper than designing and building circuits with individual components.

So what is a transistor?

We know there are millions on each chip, but what are they and what can they do? Transistors are formed from pure semiconductor material that has been carefully doped with impurities so that it either has an excess (n-type) or shortage (p-type) of electrons. Bipolar transistors are effectively a sandwich of a thin region of n-type between two p-type layers (pnp transistor) or the reverse (npn transistor). In either case, the current flowing through the device is controlled by the signal applied to the junction.

In a transistor radio, we apply a small varying signal to this junction and get an amplified version of this signal out. For many applications, especially in a digital computer, we are only interested in '0's and '1's, so the main transistor current is either on or off, controlled by a signal to the junction. Effectively it is being used as a tiny switch but, compared with other switches, it has no moving parts and can switch millions of times per second.

Field effect transistors (FETs) also depend on the junction between pnp and npn material, but this is arranged differently. They are often referred to as monopolar transistors. There are several types. They have some disadvantages compared to bipolar, but are easier to fabricate and take up less space than bipolar transistors, so can be integrated in their millions into an integrated circuit chip. They also have a very high input resistance, which makes them useful for voltage measuring devices like oscilloscopes, as they have minimal effect on the measured circuit.

How does this all work in a microprocessor?

Microprocessors are essentially carrying out the same actions as at the heart of any computer. However, they are programmed for limited and specific tasks within the circuit where they are installed. They do not have large memory systems, nor input keyboards.

There are three basic operations:

✳ calculations, such as add, subtract, multiply and divide

✳ decision making, based on information received and activation of new processes

✳ controlling the transfer of data/information from one place to another.

Activity

A microprocessor performs operations in many ways. Below is a list of parts of a microprocessor circuit. Individually you are to research the following words and write a definition of no more that 50 words for:

✳ clock line

✳ reset line

✳ address bus

✳ data bus

✳ RD & WR line

As an extension, draw a simple diagram of a microprocessor in operation and label all the parts.

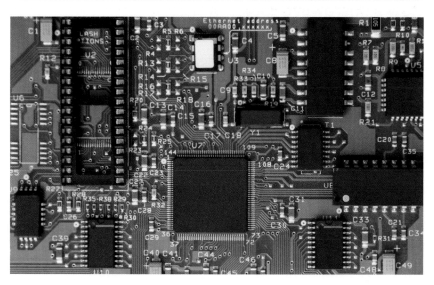

Microprocessor magnified 1000 times

Chip – very thin slice of silicon a few millimetres across, chemically etched with transistors and other components

CPU (central processing unit) – the heart of a computer that carries out computation and processing

Transistor – a semiconductor device that can amplify a small current or act as a switch

Just checking

✳ How big are microprocessors?

✳ What operations do they perform?

✳ How many transistors might there be on a single chip?

✳ What are microprocessors made from?

Microprocessors in domestic appliances

You now know that microprocessors are used on many domestic appliances. Sometimes they just drive a digital clock. In others they allow us to select and modify programmes, such as different wash cycles on a washing machine. One of the most complex is the **microwave** oven. You will look in more detail at how these work to understand what part is played by microprocessors.

How do microwaves cook?

Microwaves are a form of electromagnetic radiation that also gives us radio, television, infrared heat, light and X-rays. At microwave frequencies the energy is easily absorbed by any conducting material, where it sets up tiny eddy currents that heat the conductor up. In a microwave oven, the microwaves are absorbed by those parts of food that can conduct electricity – mainly water. (Almost all foods contain some water.) This causes the temperature of the food to rise and cooks it.

The difference from other cooking methods is that food is cooking as fast in the middle as on the outside – often faster because the microwave energy is concentrated in the centre of the oven. Other cookers cook the food from the outside in, which is slower but can seal in flavours. Some microwaves include extra radiant elements to 'brown' the outside for improved flavour and appearance.

How a microwave cooker works

The main components are shown in the table below.

Cooking food inside a microwave is effectively a closed-loop, negative feedback – system that is controlled by the timer.

Hopefully you have discovered a number of sub-systems, each controlled by the microprocessor, such as ensuring that power is not supplied until safe to do so, and that the light only works when the oven is in use, etc.

Component	Description
Key pad	To select programmes (e.g. defrost), enter desired power, timing, switch on, pause or switch off
Display panel	Displays power, time, etc. during programming and timing data during operation
Interior light	Operates when the door is open
Safety lock	Switches off power if the door is opened
Turntable	Rotates food to give more even cooking
Magnetron tube	Produces microwave energy at 2450 MHz to cook the food
High voltage transformer	Increases the input voltage from 240 V a.c. to about 5000 V d.c.
High voltage capacitor	Further raises the voltage from the transformer to start the magnetron tube
High voltage relay	Switches the magnetron tube on/off to regulate the heat
Control unit (microprocessor)	The brains behind the microwave oven, ensuring components work correctly and regulate power to the required parts

Health and safety

Microwaves will cook your hand as fast as a steak! Microwave ovens have safety locks that should switch off the power when the door is opened. NEVER put your hand (or anything living) into a microwave that is working.

Mobile phones work at similar frequencies but the energy is tiny and risks of damage are considered negligible. However, there are some concerns over long-term damage, especially to children using mobile phones a lot. There is continuing research into the effects.

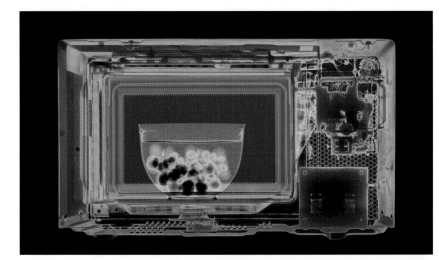

Microwave – a form of electromagnetic energy used in microwave cookers and other items like mobile phones

Microprocessor – a computer on a chip

Magnetron tube – electronic device used to generate microwaves

Internal operations of a microwave oven

Microprocessor control in petrol vehicle ignition systems

Imagine that every part of a car's **ignition** system was controlled by you. You would have to know exactly when to increase the voltage to the coil, how fast the engine is turning (RPM), and how much air is being taken in by the engine – the volume air flow (VAF). In addition to these, you would need to know exactly where the top dead centre (TDC) is for each cylinder in turn. Older cars did this mechanically, which meant that they had a lot of components that could fail and, unless the car engine was regularly and skilfully tuned, it lost power and wasted fuel. Microprocessors allow much tighter control.

The basics of a petrol vehicle ignition system

For most people, the ignition system only relates to the initial starting of the vehicle. However, it is actually the important part of the vehicle system that regulates the voltage to the spark plugs, so that combustion takes place at just the right time in each cylinder to keep the engine running smoothly.

Basically, the cylinder requires a spark to be delivered just before each piston reaches TDC, as it takes time for the spark to ignite the fuel/air mixture. Because fuel burns at a constant rate, the timing of the spark can be determined easily at 'tick over'. However, when the engine is under load (accelerating), the spark timing must be advanced. This is because the piston inside the cylinder is moving faster.

As you can see, the **engine control unit (ECU)** is connected to the coil. The spark happens at the spark plug due to the high voltage from the coil. The voltage can be 40,000 volts, even up to 100,000 volts. The high voltage is caused by the collapse of the magnetic field in the primary coil, which then induces a current in the secondary coil (which has many turns of fine wire compared to relatively few in the primary coil).

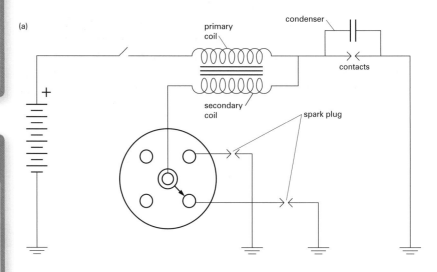

Components of a petrol ignition system

(b)

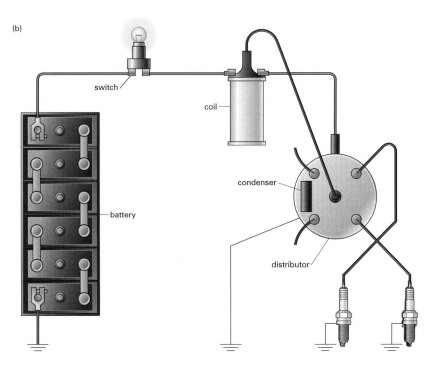

Engine control unit (ECU)

Engine control units

Modern engines have an engine control unit ECU, which contains a microprocessor. The simplest ECUs just control the quantity of fuel injected. However, the timing of the spark is crucial to the efficiency of the vehicle. In modern ignition systems, there is a sensor that tells the ECU the exact position of the pistons in the cylinders. The ECU then controls a transistor to open and close the current to the coil. The system may include several coils. Additional sensors may feed back data about the piston position, engine RPM and the volume air flow.

Other uses of microprocessor control

Whether vehicles are diesel- or petrol-driven, microprocessors are used increasingly in modern cars to help control many other functions. Think about the average family car. Are the instrument displays **digital** or **analogue**? Does the interior light go on and off just with the opening of the door (which could be a simple mechanical switch) or does it go off after a few minutes? All these are likely now to be controlled by microprocessors.

Ignition – in vehicles, causing a fuel/air mixture to burn

Engine control unit (ECU) – electronic system that controls the operation of several aspects of an internal combustion engine

Digital instrument – an instrument that takes a physical measurement, converts it electronically and displays it as a number

Analogue instrument – an instrument on which measurements are shown physically, such as a pointer on a dial

Activity

In groups of not more than three, research and list as many tasks as possible that microprocessors assist with in modern vehicle systems. Include the sensors that the microprocessor controls, with explanations on how the data is received.

Just checking

* What is an ECU on a vehicle?
* Why do we use spark plugs on petrol engines and not on diesel engines?
* What is the job of the coil on petrol engines?

What can my mobile phone do?

Make a list of the things that you do with your mobile phone: send voice messages, send text, store numbers, take photos, etc. Now go through the menus on your phone, or look at the manual that came with it, and see if there are other functions available that you do not use. Most people use very few of the features on their phones. Are they worth having?

Digital signal – a signal made up of '1's and '0's, as in binary code

Analogue signal – a signal where the sound is reproduced as a radio signal, which varies in amplitude (size) to match the change in sound

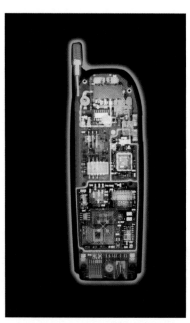

Internal diagram of a mobile phone

Microprocessor control in mobile phones

We have all seen the images of old-fashioned mobile phones, so large that they need a backpack to carry them around. Every year, phones seem to get smaller, and yet include more and more features. How is this possible? A key reason has been advances in battery technology, combined with more efficient designs requiring less battery power. However, the main reasons are the switch from analogue to digital systems and the use of increasingly powerful microprocessors.

Basic construction of a mobile phone

The electronics inside a mobile phone have not really changed much since their first introduction. Essentially the mobile phone simply converts the user's voice into a **digital signal** that is transmitted by the antenna to a base station. In the same way, the mobile phone receives a digital signal and converts it into the voice of the person on the other end.

The main components inside the common mobile phone are listed in this table.

Component	Task
Keypad	Inputting data as numbers and text
Microphone	To receive sounds, voice or music
Speaker	To play sounds, voice or music
Antenna (usually internal)	To send and receive the radio frequency pulses
SIM card	Subscriber Identity Module, able to store data, such as details of contacts, telephone numbers, as well as calendar information
SRAM	The area of memory that holds the operator's information
Dual band RF module	Receives the digital signal from the analogue to digital converter, and converts it into a radio frequency pulse
LCD display	To display information
Analogue to digital converter	To transfer sound communication into digital so that it can be transmitted over the network; and convert digital signals into analogue voice data for received signals
Microprocessor	To control everything above

Activity

You should already have listed the functions on your own mobile phone. Now add to that list by researching at least six of the latest phones advertised on the internet. Construct a spreadsheet with a column for each phone. Enter the identified functions as rows and put a '1' or '0' in the appropriate cell to indicate whether each phone has the function or not. Total these columns to show which phones have the most functions.

Discuss your results with others in your class and see if there are any other functions that you could add.

As an extension, make a copy of this spreadsheet and use the new version to grade the phones on the value of their functions. For example, you could decide that an 'essential' function was worth 5, a function that was 'useful but not essential' was worth 3 and a function 'of little use' was worth only 1. (Add some extra categories or weight the values differently, if you wish.) Enter the selected value in each cell wherever there is a 1 to indicate that the phone has that function. The column totals should now give a much better comparison of the real worth of each phone to you. If others in your class have done this, do you agree on which is the best phone and, if not, why not?

Activity

As you can see from the image, the mobile phone has a lot of electronic components, all of them in some way connected to the microprocessor. On a sheet of A4, draw how you think all the parts are connected. Make your drawings clear and legible, as this will form part of your work to show that you have understood this topic.

Operation

Mobile phones operate over a network of cells, each with a base station antenna at its centre, which are placed across the country. There may be several networks in each country (e.g. Vodafone or Orange), but they can talk to each other and to fixed line telephones.

Each mobile phone has a unique identity, called the International Mobile Equipment Number (IMEI), built into its own microprocessor memory. In addition, when you as a subscriber sign a contract to be able to operate on one of the networks, a SIM card is placed in the phone with your unique International Mobile Subscriber Number (IMSI). Whenever your phone is switched on it contacts the nearest base station, which registers its presence within the cell. If you move into another cell area, your new location is noted. If you make a call, the network finds the location of the phone you are calling, whether on the same network or any other, and connects the phones.

The first generation (1G) phones were analogue. 2G (and later) phones are digital, allowing increasing facilities and security. For these, when the microprocessor registers sound through the microphone, it sends it through to the analogue digital converter. Once the sound has been converted, the microprocessor sends the digital data to the RF module where it is converted to a radio frequency for the antenna to send.

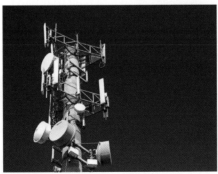

Mobile phone base station mast

Functional skills

Do some research in the library or on the internet to find out what type of signal is transmitted across the network. How is this signal achieved? Produce a short report, including diagrams, to describe your findings.

As an extension, produce a report of not more than 250 words explaining the difference between 2G and 3G mobile technology.

Just checking

✱ Why are microprocessors used in mobile phones?
✱ What is an IMSI and where would it be stored on a phone?
✱ What is an IMEI and where would it be stored on a phone?

Vehicle diagnostic systems

Imagine a car that not only discovered and recorded faults, but also managed to repair itself. Could science fiction come true? The automotive industry is halfway there. Vehicles today are fitted with sophisticated electronic networks that make traditional automotive repair methods difficult to apply. Vehicle diagnostic tools are essential, even for the simplest of **maintenance** operations.

Where and when fitted

Most modern cars, manufactured after 2001, will be fitted with at least one diagnostic slot. This enables the engineer to plug a diagnostic tool into the slot and receive data from the vehicle's ECU (engine control unit) and other sensors on the vehicle. As discussed earlier, ECUs contain microprocessors, which are used to monitor and control a number of engine functions. There may then be other sensors, looking at everything from brake wear to tyre pressures, which may send their data to the ECU or to other microprocessors on the vehicle.

How does it work?

The diagnostic tool is a computer – often a laptop or hand-held device – which is loaded with all the data for the particular vehicle. The data is usually provided on a CD-ROM, which the manufacturer will update as necessary and send out to all garages equipped to repair that type of vehicle. The data would then be downloaded, if the tool was a handheld device, or accessed directly, if using a laptop with a CD reader.

How does plugging the tool into a slot help with diagnosing a fault within the vehicle? With earlier systems, once the device had been connected, it displayed a number or code. This code related to a fault that had been noted and recorded inside the ECU. Once the code had been found in the diagnostic manual, repair or maintenance could take place. With increasing computer power and memory,

Diagnostic progression

the diagnostic device can now give access to all the relevant information, including instructions on how to conduct repairs and of parts to be replaced.

At the same time, the computer will usually interrogate the other systems on the vehicle and provide information for analysis by the engineer. This could indicate faults that are likely to occur soon, such as brake discs needing replacement.

Latest developments

The images show just how quickly diagnostic tools are developing. In fact, they have already gone several steps further. Microprocessors on vehicles are being fitted with communication systems, so that faults can be reported directly to a base station, though this is principally on large vehicles. Complete commercial vehicle fleets are being operated with a base station continually monitoring not just the location of a vehicle but also engine performance, tyre pressures and even how efficiently the driver is driving. These operate over radio nets or use automatically generated mobile phone messages.

On cars, some microprocessors and diagnostic tools have been fitted with **Bluetooth™**. These let diagnostic tools check vehicle performance without having to be plugged into the vehicle, and can be used during a test drive. As an owner of the vehicle, you can see up-to-date information as it is received. In the case of a software fault providing an incorrect error, you are able to clear the coding, which some servicing centres charge for.

Microprocessors cannot repair damaged parts but can control how some elements in the vehicle system perform. Being able to communicate with the vehicle microprocessors means that adjustments can be made through the diagnostic system. These can cover such items as suspension, breaking, traction control, the fuel system, plus many more.

In many cases, the automotive engineer will be able to rectify faults without the need for any additional tools other than the diagnostic device. Tweaking or resetting to factory settings can solve many faults.

Personal learning and thinking skills

Apart from recording and displaying fault codes and information, what else can a diagnostic tool do? In pairs, think about owning a device that enabled you to communicate with your vehicle. What do you think you would be able to do with it? As a class, discuss the possibilities. Your teacher will produce a list from your ideas.

Maintenance – activity involved in keeping something in working order
Bluetooth™ – wireless short-range communication system

Personal learning and thinking skills

List as many microprocessor systems as you can find for the modern vehicle. Your teacher will create a list and display these on the board.

Activity

As you have discovered, the automotive industry commonly uses diagnostic tools. Use the library or internet to research other industries that also use such diagnostic tools. Produce a report of no more than 250 words that covers the type of diagnostic tools used, where they are used and for what purposes. Include any problems that you find reported in their use and ways these are being solved.

Just checking

* What can engineers use as diagnostic tools?
* How do manufacturers ensure that vehicle repair and maintenance engineers are kept up to date on changes to vehicle systems?
* When did the inclusion of diagnostic systems on most vehicles become standard?

Personal learning and thinking skills

By yourself, write a list of at least five operations or tasks for which a terminal could aid an engineer. Your teacher will create a list of all the answers given for discussion by the class. Make notes during the discussion of other ideas that you think could be of value and write up the extended list.

Analyse – to investigate and explore to determine a result

Database – a collection of data arranged for ease and speed of search and retrieval

Bar code – a numerical code that can be printed on objects, able to be read by a scanner and interpreted by a computer

Hand-held computers

As technology within engineering increases, so too does the maintenance requirement. The wealth of information needed by any engineer, whose job it is to maintain, **analyse** or use the systems, is immense. Ideally, engineers need to be mobile, efficient and paperless. In the past, technical manuals would be needed for any maintenance job. Nowadays, with a laptop or hand-held computer, engineers can just scroll through the **database**, select the model and open the PDF file – information at their fingertips!

Why use a hand-held computer rather than a laptop?

Laptop computers have become lighter and more powerful every year. Away from the office, they can provide almost all the facilities of a desktop computer and can be connected up, via a mobile phone or landline, so that they are effectively an extension of the office network. They are vital to staff like salesmen or engineers needing access to plans or large documents. However, they have to be placed on a surface to use them and need two hands to operate. They do not always stand up to rough handling, or operating in adverse weather conditions.

Many engineers want computer support for specific tasks only. Often they want something small, robust, able to be operated with one hand and with a limited screen and keyboard. Hand-held computers (often just referred to as 'terminals') fill this need. They can be designed for specific tasks and environments and, yet, can still be linked to more powerful computers on a company network, or even access the World Wide Web. Hence, users can still enter and transmit data, request and read information, or search for it in a database.

Wireless technology

Bluetooth™ systems now allow hand-held computers to link into networks within a local area, such as a factory, office block or warehouse. Other wireless technologies are available (e.g. radio nets, Wi-Fi, GPRS, 3G mobile phone links) that allow hand-held computers to be linked over unlimited distances.

Using a hand-held computer in a warehouse

Two main objectives of companies are to reduce overheads and cut costs. One method is to limit the amount of storage needed for raw materials and processed products. The hand-held computer is an ideal tool for the job. Using wireless technology, it can instantly send requests for materials needed, as well as inform the dispatch department that orders have been completed. For instance, when staff walk around the warehouse they may notice that some stock is low. Using a terminal designed for the task, the **bar code** for the specific stock can scanned and the information instantly transmitted to the suppliers. This can indicate the quantity remaining and how soon the stock needs to be replenished.

Activity	Technology
An engineer walks around the warehouse and notices the stock of raw materials is low. Using a terminal, she scans the barcode on the allocated stock bay.	Bar code reader
She connects to the internet	WiFi
She logs on to the supplier's secure net access	WWW
She uploads the request on to their system	Database
The system instantly sends a text back with a day for delivery	SMS

Hand-held bar code reader

Flexible network

Activity

Your task is to develop further the operation you selected earlier. Using the table above as an example, you need to find images for your operation and create a poster outlining the terminal interactions.

As an extension, for each of the interactions you need to write exactly what processes are taking place and the technologies used.

Functional skills

Use of a hand-held computer

From your earlier list, select one of the operations that a hand-held computer can aid with. Write in detail exactly what stages and people are involved, including the technologies used. An example is illustrated in the table.

Just checking

* What other name is commonly used for hand-held computers?
* How can they communicate?
* Where would you find a bar code?

Vision systems

We can see this text, read it and understand it, because our brain knows how to interpret the information. However, it took most of us, as children, two to three years to fully recognise what letters and words mean. How could a computer see, read and understand? You covered this briefly in Topic 5, where you looked at the problems computers face with pattern recognition. Here you will look more closely at vision systems that are used in such areas as production lines.

Types of vision sensors

We have all seen the CCTV cameras in operation in our streets and local shops. These primarily record information, though some are manned and the images are watched live. If CCTV operators observe a potential problem, they notify the authorities who deal with it. With 'smart vision systems' and 'smart cameras' on production lines, there is no one watching the images being received. Instead, the image is interpreted by the system and, depending on what is observed, certain actions are carried out. If the image being received by the cameras is what is expected, production continues as normal. When the images do not conform to expectations, a signal is sent to carry out some action. Ultimately the action is likely to be rejection of the product or component being viewed. In its simplest form, it may just sound an alarm to attract an operator's attention and/or stop the production line. Alternatively, it may trigger some automated way to remove the incorrect item from the line without interrupting production.

Why use a vision system during production?

Reasons could include:

* enabling faster operation than just using manual monitoring
* generation of comprehensive production statistics to monitor performance
* improving yield.

Use of smart vision sensors on a production line

Smart vision sensors can be used for a number of applications. However, because they often only come with simple **algorithms**, they are best suited to tasks that need to know just whether something is present or absent. They have been developed further to support specific manufacturing, such as:

* plastic sealing, for applications such as vacuum food-processes packaging, to detect the effectiveness of the seal
* crate inspection, to ensure that packaging crates are free from deformation prior to filling with products
* aluminium foil production
* checking that metal cans are properly sealed after filling
* print control, to check for legibility, clarity, etc.

* accuracy and placing of labels on products
* box inspection, once products have been packaged and sealed, to ensure that there has been no damage and that seals are in place.

Two different types of vision system

Both of the vision systems shown in these photos need the data to be interpreted. In the case of the human eye, it is the brain that identifies the image from its characteristics, whether these be font, colour, shape or even texture. For the electronic vision system, it is a computer. This will have a number of algorithms and image files stored, against which the received data is checked.

If the vision system is being used on a production line to identify errors, then, if there is a match, the process continues without interruption. However, if the received data from the vision sensor does not match the expected image, then the system may halt production until engineers inspect and diagnose the fault.

Unit 3 Assessment Guide

In this unit, you will investigate how engineering industries use computer technology in their manufacturing systems, products and services so that they remain competitive. You will also investigate how microprocessors are used to control the operation of everyday consumer products and the application of computer technology to support maintenance operations.

Time Management

Manage your time well as this unit has a number of different components that will have to be researched. Ensure that you keep your work safe and that any work in electronic format has a secure and safe backup.

Be well organised. This is your chance to show that you are an independent enquirer, creative thinker, reflective learner and self manager and therefore will contribute towards achievement of your Personal learning and thinking skills.

Be prepared with a list of relevant questions that you could ask if your teacher arranges a visiting speaker or a visit to a business which uses computer diagnostic equipment such as the service department of a garage.

Plan ahead for your work experience and make a list of things that you need to find out or observe so that you make the most efficient use of your time.

Useful Links

Make good use of your work experience to find out as much as possible about issues that are relevant to your coursework. You may have the opportunity to work and meet with people who operate machinery and equipment which is controlled by computer systems and microprocessors.

Things you might need

Your work needs to be in the form of an A4 word-processed report which contains digital images showing you carrying out practical work. It should be presented as an e-portfolio and your teacher will give you access to the required software to enable the correct presentation.

Your teacher should give you access to computer-aided equipment such as a rapid prototyping machine.

You would benefit from access to a technician who works with computers which control processes or machinery.

A digital camera or mobile phone would be useful, so that you record evidence of practical activities.

Remember to maintain a focus on why computers and microprocessors are used to control the operation of processes, manufacturing systems and consumer products.

How you will be assessed		
What you must show that you know	**Guidance**	**To gain higher marks**
How computers are used to control the processing of products (e.g. filling bags of crisps) and the manufacturing of components (e.g. cutting metal using a machining centre). *Assessment Focus 3.1*	✳ Investigate how a computer system is used to control a flow process e.g. running a packing line or monitoring quality. ✳ Investigate how a computer system is used to control a manufacturing process e.g. a robot welding car bodies.	✳ You need to compare the two systems and explain the differences in the way that they function. ✳ You need to evaluate the effectiveness of using computer control for each of the systems investigated.
How to solve an engineering problem using computer-based equipment as a tool. *Assessment Focus 3.2*	✳ Your teacher will give you an engineering problem to solve e.g. checking the dimensions of a machined component using a co-ordinate measuring machine (CMM). ✳ Set up and use computer based equipment and make a record of what you did. ✳ Obtain witness statements and take photographs to prove that you carried out the practical work effectively.	✳ You should identify any safety issues which relate to using the equipment. ✳ You need to justify why computer based equipment was chosen rather than non-computer based equipment for solving the problem. ✳ You need to appraise the effectiveness of the method used to solve the problem.
How microprocessors are used to control the operation of many types of everyday consumer product. *Assessment Focus 3.3*	✳ Identify two different consumer products which are fitted with microprocessor control systems. ✳ For each product, describe how the control system is used to make it operate. ✳ Identify the major components of each control system.	✳ For just one of the products you should evaluate the effectiveness of its microprocessor control system. ✳ You need to explain how the component parts of the control system work. ✳ You need to explain how the control system could be used in a different product.
How computers are used to improve the way that maintenance operations are carried out. *Assessment Focus 3.4*	✳ Identify two different maintenance operations which require the use of computer aided technology e.g. a diagnostic check on the electronic control unit (ECU) of a car engine. ✳ For each application, describe how the computer technology is used.	✳ You need to explain how computers enable detailed fault analysis to be carried out. ✳ You need to gather computer data generated from a maintenance operation. ✳ You need to interpret the data, evaluate it against a specification and propose a course of action should it be necessary.

4 PRODUCING ENGINEERING SOLUTIONS

Introduction

This unit will provide you with an insight into how engineers go about solving problems.

It will also be an opportunity to experience some of the skills and processes that are used in the production of an engineered product or the delivery of an engineering service. This might involve the maintenance, installation or commissioning of an engineered product.

Planning is a crucial skill in engineering. When you study this unit, you will need to have an effective plan that identifies each stage in the production/delivery process, and which takes into account the materials, resources, tools and processes needed.

Being able to select an appropriate material for a given engineering task is another important skill that engineers must have. You will need to be able to justify your selection of materials, based on a variety of considerations, such as strength, weight, durability, corrosion resistance and cost. You will also need to be able to identify and use a wide variety of electrical and mechanical components and parts, such as resistors, transformers, motors, drives and gears. Being able to make effective use these of these 'off the shelf' components and assembling them into a working product, is important in a huge variety of engineering applications.

When you have completed your engineered product you will need to know that it has been manufactured and assembled correctly, and that it performs in the way you expect. To do this you will need to be able to apply appropriate inspection techniques used by engineers to confirm that a product is fit for purpose and is ready for service.

This unit provides you with an exciting opportunity to put your ideas into practice and apply a range of practical engineering skills. It is, therefore, important at the outset that you fully understand and can apply appropriate health and safety procedures, and that you know how to carry out a risk assessment for an engineering activity. You need to remember that safety, both your own as well as that of others around you, should be paramount in all of your practical engineering work.

This unit is assessed by your tutor. On completing it you should:

1. Understand health and safety procedures, standards, and risk assessment in engineering activities.

2. Be able to plan for an engineering product or service.

3. Be able to select suitable materials, parts or components for an engineered product or service.

4. Be able to use processes, tools and equipment to make an engineered product or carry out a service.

5. Be able to apply inspection techniques to the engineered product or service.

THINKING POINTS

Think about the following key points as you work through this unit:

1. Why is Health and Safety important when performing an engineering task or delivering an engineering service?

2. Which engineering processes and which engineering materials are hazardous, what makes them hazardous, and how can the risks associated with using them be minimised?

3. Why is it important to do a Risk Assessment before carrying out an engineering activity?

4. Why is planning important in an engineering context and what is the purpose of a job instruction or work sheet?

5. Why are particular materials, tools and processes chosen for a particular engineering application?

6. What are the main types of electrical and mechanical component available to engineers, and how would you go about selecting a component for a particular application?

7. How would you check that a finished product or engineering service complies with specification and is fit for purpose?

Health and safety

Many engineering processes are potentially hazardous and these include activities such as casting, cutting, soldering and welding. In addition, some processes involve the use of hazardous materials and chemicals. Even the most basic and straightforward activities can potentially be dangerous if carried out using inappropriate tools, materials, and methods.

Hazards and hazardous processes

Engineering processes that are particularly hazardous include:

* casting, forging and grinding
* chemical etching
* soldering, welding and brazing
* heat treatment
* cutting and forming
* use of compressed air.

In all cases, the correct tools and **personal protective equipment** should be used and proper training should be provided. In addition, safety warnings and notices should be prominently placed in the workplace and access to areas where hazardous processes take place should be restricted and carefully controlled-only appropriately trained personnel should be present. Storing hazardous materials (chemicals, radioactive substances, etc.) requires both special consideration and effective access control.

While engaged in engineering activities, you need to observe a number of health and safety precautions such as:

* follow safety guidance and approved procedures for carrying out the work (your teacher will provide you with specific guidance on this)
* make proper use of equipment provided for your safety
* inform your teacher (and other appropriate persons) if you become aware of anything that could be hazardous
* take care to ensure that your activities do not put any other people at risk.

A grinding machine

Hazard – a condition that constitutes a risk to health and safety

Personal protective equipment (PPE) – clothing, footwear and other items that can be worn or carried to avoid risks and minimise hazards

Tools and tidiness

A sign of a good engineer is a clean and tidy work area. Only the minimum of tools for the job should be laid out at any one time. These should be organised in a logical manner so that they immediately fall to hand. Tools not immediately required should be cleaned and properly stored away. Hand tools should be regularly checked and kept in good condition. Spillages, either on the workbench or floor, should always be cleaned up immediately.

Human carelessness

Most accidents are caused by human carelessness or negligence. This can range from 'couldn't care less' and 'macho attitudes', to the deliberate disregard of safety regulations and codes of practice. Carelessness can also result from fatigue and ill health and these, in turn, can result from a poor working environment.

Personal habits

Personal habits, such as alcohol and drug abuse, can render workers a hazard to themselves and others. Fatigue due to a second job ('moonlighting') can be a considerable hazard, particularly when operating machines. Smoking in prohibited areas where flammable substances are used and stored can cause fatal accidents involving explosions and fire.

Supervision and training

Other causes of accidents are lack of training and inadequate or poor quality training. Lack of supervision can also lead to accidents, if it leads to safety procedures being disregarded.

Environment

Unguarded and badly maintained plant and equipment are obvious causes of injury. However, the most common causes of accidents are slippery floors, poorly maintained stairways, scaffolding and obstructed passageways in overcrowded workplaces. Excessive noise, bad lighting and inadequate ventilation can lead to fatigue, ill health and carelessness. Dirty surroundings and inadequate toilet and washing facilities cause sickness.

Elimination of hazards

The workplace should be tidy, with clearly defined passageways. It should be well lit and ventilated. It should have well maintained non-slip flooring. Noise should be kept down to acceptable levels. Hazardous processes should be replaced with less dangerous and more environmentally acceptable alternatives. For example, asbestos clutch and brake linings should be replaced with safer materials.

Guards

Rotating machinery, drive belts and rotating cutters must be securely fenced to prevent accidental contact.

Some machines have interlocked guards. These are guards coupled to the machine drive in such a way that the machine cannot be operated when the guard is open for loading and unloading the work.

All guards must be set, checked and maintained by qualified and certificated staff. They must not be removed or tampered with by operators.

(a) Guard fitted to a pillar drill

(b) Guard fitted to a cutter

(c) Guard fitted to the drive belt of a power guillotine

(d) Guard fitted to the revolving bar of a lathe

Revolving bar

Barrier

Machine guards

Personal learning and thinking skills

Visit the Health and Safety Executive (HSE) website at www. hse.gov.uk and obtain information on safety in the gas welding and cutting process. Identify at least FIVE main hazards associated with the process and explain the meaning of 'backfire' and a 'flashback'. Present your answer in the form of an A3 poster than can be displayed in the welding area of an engineering workshop.

As well as processes, some materials used in engineering can be hazardous. These include many fluids, such as hydraulic fluid, etching fluids, battery acid, etc. All hazardous materials should have a Material Safety Data Sheet (MSDS) from the material manufacturer or supplier. The sheet should be identified by the description and part number of the product and should contain comprehensive information about the product, including details relating to handling, use and transportation. Details of protective clothing or equipment required to work safely with the product should also be included. If you are working with hazardous materials, you should have access to the MSDS for the materials concerned.

Working safely

As you progress in your engineering course, you will be expected to show that you can work efficiently and effectively in an engineering environment. In this unit, you will plan and carry out an engineering activity. To do this safely, you will need to observe the following safety precautions:

* **Prepare the work environment** – Ensure that it is free from hazards and that relevant safety procedures are implemented. You also need to ensure that appropriate PPE and tools are selected and checked before use, and that each is in a safe and usable condition.

* **Prepare the work activity** – Ensure that all necessary drawings, specifications, job instructions, materials and component parts are obtained, and appropriate storage is available. You must also ensure that you have the necessary authorisation to carry out work from your teacher and/or workshop supervisor.

* **Complete the work activity** – Include all specified tasks and associated documentation, paying particular care to correct working practices and observing any safety guidance. When completed, you should return drawings, instructions and tools to the correct storage location and dispose of unusable tools, equipment, components and waste materials (oil, soiled rags, swarf, off-cuts, etc.).

Personal protective equipment

Personal protective equipment (PPE) must be used when performing many engineering tasks. It includes clothing and footwear as well as specialist equipment for eye protection, ear protection, breathing apparatus, etc.

An extract from a Material Safety Data Sheet

✳ **Clothing and footwear** – Suitable and unsuitable clothing for use in an engineering workshop is shown in the diagram. Overalls or protective coats should be neatly buttoned and sleeves should be tightly rolled. Safety shoes and boots should be worn (not trainers!). Overalls and protective clothing should be sufficiently loose to allow easy body movement, but not so loose that they interfere with engineering tasks and activities.

✳ **Special equipment** – Some processes and working conditions demand even greater personal protection. In such cases, PPE can include safety helmets, ear protection, respirators and eye protection, worn singly or in combination. Such protective clothing must be provided by the employer when a process demands its use. Employees must, by law, make use of such equipment.

a Wear the corrdct protective clothing

b Wear safety headgear

c Wear safety shoes or boots

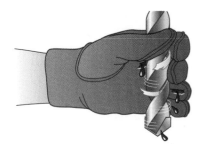

d Wear gloves when handling sharp objects

e Wear eye protection when using machinery

f Wear a respirator when dust and fumes are present

Examples of personal protective equipment (PPE) and when to wear it

Risk assessment

It is now a legal requirement for engineering companies to carry out a systematic assessment of the risks associated with their activities. The purpose is to ensure that risks are reduced as far as is reasonably practicable. Risk assessment must also be carried out when changes are made to production processes, such as the introduction of different tooling or materials. Risk assessment is something that you will almost certainly become involved with at some stage in your career, so it's important to know how it's done.

The risk assessment process

Engineering companies are required to carry out regular risk assessments to identify hazardous activities and to inform action plans designed to improve safety and eliminate hazards. Correct storage and handling of materials is essential. Personal protective equipment (gloves, goggles, overalls, etc), correct lighting and workspace organisation, adequate ventilation and efficient fume and dust extraction all have a part to play in making the workplace safe.

The **level of risk** is important and is determined by the probability of an event and the likely severity of its outcome. Evaluation of the risk involved in a production process or activity is usually based on the answers to simple questions, such as these:

* What are the hazards?
* How likely are they to occur?
* Under what circumstances might they occur?
* What controls are in place to prevent them occurring?
* What are the likely consequences if they do occur?

The process of identifying hazards can be a little daunting, if you have never done anything like this before. To make things a little more manageable, you can divide the task into a number of smaller questions:

What

– engineering processes are being used and are any hazardous?

– tools, equipment, materials and PPE are being used?

– can go wrong and how can it be prevented?

– are the statutory requirements and how must I comply with them?

Who

– is undertaking the task and what training/experience have they had?

– might be exposed to risk and who else might be affected?

– is responsible for checking/supervising the task?

Where

– is the task being performed?

– is the task recorded and who needs to keep the record?

– do waste materials go?

When

– is the task being carried out?

– will the task next be carried out?

– were the safety procedures last reviewed/updated?

Why

– is the task being carried out?

– is the task performed in the way that it is?

– must PPE be used?

How

– could an accident occur?

– can the risks/hazards be controlled?

– can the level of risk be assessed?

Activity

Perform a risk assessment of an engineering task, such as cutting, bending and drilling sheet metal, taking into account correct use and storage of personal protective equipment, health and safety regulations, and warning signs. Ensure that your risk assessment includes the hazards present, who might get harmed, the risks, control measures and precautions that need to be observed.

How to carry out a risk assessment

The risk assessment process is more effective if it is carried out by at least two people – the person who will carry out the activity and/or is familiar with the process and the person responsible for the activity. This ensures that the risk assessment is better informed and is more objective. It also helps to ensure that there is 'ownership' of the risk assessment.

The steps involved

* Describe the task and the circumstances and environment in which it is performed.
* Identify and list the hazards associated with the task.
* Estimate the risk arising from each of these hazards.
* Who is likely to be affected and what is their understanding of the risks?
* How effective are existing control measures (e.g. safety guards, barriers, etc.)?
* Are any additional control measures required?
* How effective is the PPE provided (overalls, boots, eye protection, hard hats, etc.)?
* How will you communicate the contents of the risk assessment to those who may be affected? .
* How often will you need to review the risk assessment to ensure that it remains current and effective?

Personal learning and thinking skills

A local engineering company has set up a new manufacturing plant, which will incorporate a gas welding facility. The company has asked you to advise them on the hazards associated with gas welding. Prepare a ten minute presentation using appropriate visual aids (e.g. PowerPoint slides or a flip chart) to explain the hazards, relevant health and safety requirements and the need for a formal risk assessment of the new facility.

Just checking

* What is a risk assessment and why is it important?
* How should a risk assessment be conducted and what should it involve?
* What is level of risk and how is it determined?
* What is a maintenance procedure and what should be included in it?

Activity

Prepare a detailed production plan for the simple LED flasher shown in the photographs. Think about the materials, parts and components to be used and whether any of these can be bought in or need to be manufactured. Think about the tools and equipment that will be needed and how the product will be finished. Suggest a sequence and timescale for each activity. Present your production plan as a series of work instructions (including notes and sketches where relevant).

Production plan – a detailed strategy for producing an engineered product or delivering an engineering service that describes the sequence of operations, processes, materials and resources used

Work instruction/job sheet – a description of a particular operation or task that specifies what should be done, how it should be done and what materials, processes and tools should be used

Planning

Planning is an important task that needs to be carried out in detail prior to the production of an engineered product or the delivery of an engineering service. An effective plan will save both time and money. Planning involves thinking through the sequence of jobs and activities and specifying the component parts, materials and processes.

The planning process

It is essential to have a detailed strategy before carrying out an engineering activity. We call this a **production plan**. The plan is usually communicated to other people as a series of **work instructions** or **job sheets**. These specify the tasks to be performed, how they are to be carried out, the processes and tools to use, and how work is to be checked for quality.

Factors to be considered

Production planning must consider a number of factors including:

* available resources
* materials, parts and components to be used
* processes to be used
* available equipment and machinery
* sequence of production
* arrangements for inspection and quality control
* health and safety (H&S) factors.

Constructing a simple electronic product

Let's assume that you have decided to construct a prototype battery-powered alarm that will emit a sound from a piezoelectric transducer whenever a bike is moved. The alarm is to be set (made active) and reset (deactivated) using a simple two-position keyswitch.

The movement of the bike is to be detected by means of a small motion detector. This electronic component is to be mounted, with the rest of the alarm circuit components, on a small printed circuit board (PCB).

The alarm circuit is to be powered from a rechargeable battery. The assembly is to be enclosed in a sealed (waterproof) diecast aluminium alloy box, secured to the frame by a bracket with anti-tamper bolts.

The table shows the main components.

piezoelectric transducer
keyswitch
motion detector
rechargeable battery
battery holder
printed circuit board *
diecast enclosure
clamp assembly.*

You intend to manufacture the components marked * and buy in all the other parts, as they are available 'off-the-shelf'.

Since the PCB will provide a means of both mounting the electronic component and connecting them together, you will probably need to design and manufacture this first. Before you can assemble the PCB and the other components inside the case, you will need to drill the diecast box so that the PCB, keyswitch and battery holder can be fitted to it.

Deciding on a sequence of operations

Deciding on the precise sequence of operations to manufacture an engineered product can be complex, especially when a large number of components or processes (or both) are involved. Fortunately, the production sequence for your bike alarm should be fairly straightforward:

* Design and manufacture the printed circuit board.

* Obtain electronic parts and solder them into the printed circuit board.

* Drill the diecast box for the printed circuit board mounting pillars, keyswitch, piezoelectric transducer, battery holder, and clamping assembly.

* Assemble the printed circuit board and other parts in the diecast box; solder any interconnecting wires required.

* Charge the battery and insert this into the battery holder.

* Test the alarm and check that it operates correctly.

* Manufacture the clamp assembly.

* Attach the clamp assembly to the diecast box.

We have to consider the components and materials, as well as the processes and equipment.

* Why use a diecast aluminium enclosure?

* Why use a rechargeable battery?

* Why would you want to use anti-tamper bolts in the clamp assembly?

* What tools and equipment will you need to manufacture the PCB?

All these questions and more need to be answered before you start to manufacture your product.

Components and parts used for the simple LED flasher

Completed printed circuit assembly

Internal arrangement of the LED flasher

The completed LED flasher

Just checking

* What is a production plan and why is it important?

* What is a work instruction (or job sheet) and why is it important?

Spot the material!

Think about something that you use every day, such as a bike or computer. How many different materials are used in it and how many of them can you identify? Working with two or three other students, make a list of at least ten commonly used materials in engineering. For each material that you have identified, list at least two important properties that make them useful in engineering.

Selecting engineering materials

A wide range of materials, parts and components is used in modern engineering. Metals, plastics, ceramics and composite materials offer different characteristics that make them suitable for widely different applications. Being able to select a suitable material for a given application is an important task for an engineer. More information on the different types of engineering material can be found in Unit 8. Here you shall concentrate on the properties of the different types of material and how these determine the use to which each material is put.

Types of material

The chart summarises the main groups of engineering materials. The Latin name for iron is ferrum, so it is not surprising that ferrous metals and alloys are all based on the metal iron. Alloys consist of two or more metals, or metals and non-metals that have been brought together as compounds or solid solutions to produce a metallic material with special properties. For example an alloy of iron, carbon, nickel and chromium is stainless steel. This is a corrosion resistant ferrous alloy. The remaining metals are non-ferrous metals and alloys, such as copper, brass and zinc. Non-metals can be natural, such as rubber, or they can be synthetic such as the plastic compound PVC.

Corrosion – chemical attack on metals and metal alloy (for example, rust)

Degradation – gradual weakening of non-metals due to chemical attack, exposure to sunlight or other radiation

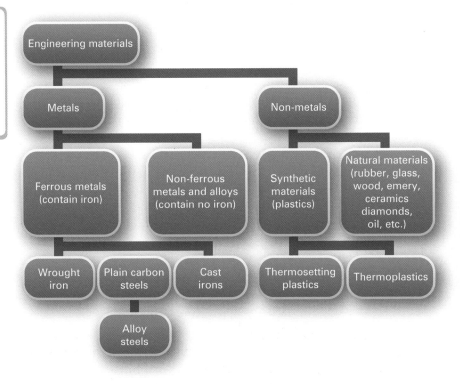

The main types of engineering material

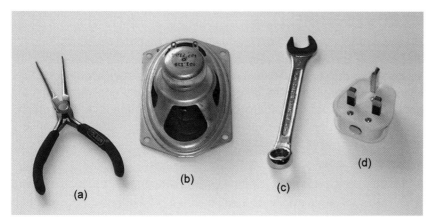

Some common engineered products that use different materials in their manufacture: (a) a pair of long-nosed pliers, (b) an elliptical moving coil loudspeaker, (c) a spanner with ring and open ends, (d) a standard UK mains plug

Properties of materials

When selecting a material, you must ensure that it has suitable properties for the job it has to do.

* ✱ Will it corrode in its working environment?
* ✱ Will it weaken or melt in a hot environment?
* ✱ Will it break under normal working conditions?
* ✱ Can it be easily cast, formed or cut to shape?

You assess the suitability of different materials for a particular engineering application by comparing their properties.

Chemical properties

✱ **Corrosion** – This is caused by metals and metal alloys being attacked by chemical substances. For example, rusting of ferrous metals and alloys is caused by the action of atmospheric oxygen in the presence of water. Another example is the attack on aluminium and some of its alloys by strong alkali solutions. Take care when using degreasing agents on such metals. Copper and copper-based alloys are stained and corroded by the active sulfur and chlorine products found in some heavy duty cutting lubricants. Choose your cutting lubricant with care when machining such materials.

✱ **Degradation** – Non-metallic materials do not corrode but can be attacked by chemical substances. This weakens or even destroys the material. Unless specially compounded, rubber is attacked by prolonged exposure to oil. Synthetic (plastic) materials can be softened by chemical solvents. Exposure to the ultraviolet rays of sunlight can weaken (perish) rubbers and plastics unless they contain compounds that filter out such rays.

Surface coatings are often applied to improve the corrosion resistance of materials. These protective coatings form a barrier between the metal and the chemicals that would attack it. Various types of coating are available, including organic types (such as polymer paints), ceramic coatings (such as enamels), metal coatings (such as tinplate) and chemically deposited coatings (such as phosphates). Phosphate coatings are often used prior to the application of paint where they act as a keying layer.

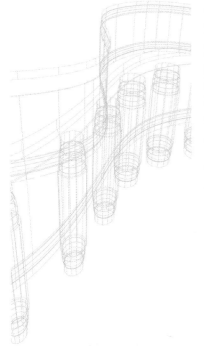

Electrical properties

Electrical properties relate to how well a given material conducts electricity. Materials with a very low resistance to electric current flow are good conductors. Materials with a very high resistance are good insulators. Generally, metals are good conductors and non-metals are good insulators (poor conductors). A notable exception is carbon, which conducts electricity despite being a non-metal. The electrical resistance of a metal conductor depends on:

* length (the longer it is, the greater its resistance)
* thickness (the thicker it is, the lower its resistance)
* temperature (the higher its temperature, the greater its resistance, except for carbon where it decreases with temperature)
* resistivity (the resistance measured between the opposite faces of a metre cube of the material).

Note that a small number of non-metallic materials, such as silicon, have atomic structures falling between those of electrical conductors and insulators. These are called semiconductors and are used for making solid state devices, such as transistors.

Magnetic properties

All materials respond to strong magnetic fields to some extent. Only the ferromagnetic materials respond sufficiently to be of interest. The more important ferromagnetic materials are the metals iron, nickel and cobalt. Magnetic materials are divided into two types: soft and hard.

Soft magnetic materials, such as soft iron, can be magnetised by placing them in a magnetic field. They cease to be magnetised as soon as the field is removed.

Hard magnetic materials, such as high-carbon steel, hardened by cooling it rapidly (quenching) from red heat, also become magnetised when placed in a magnetic field. Hard magnetic materials retain their magnetism when the field is removed, becoming permanent magnets.

Permanent magnets can be made more powerful for a given size by adding cobalt to steel to make an alloy. Soft magnetic materials can be made more efficient by adding silicon or nickel to pure iron. Silicon iron alloys are used for the rotor and stator cores of electric motors and generators, and for power transformers cores.

Thermal properties

Thermal properties relate to how a material responds to heat and different temperatures.

* **Melting temperature** – This is the temperature at which a material loses its solid properties. Most plastic materials and all metals become soft and eventually melt. Note that some plastics do not soften when heated: they only become charred and are destroyed.

* **Thermal conductivity** – This is the ease with which materials conduct heat. Metals are good conductors of heat. Non metals are poor conductors of heat. Therefore, non metals are heat insulators.

* **Expansion** – Metals expand appreciably when heated and contract again when cooled. They are said to have high coefficients of linear expansion.

Non-metals expand to a lesser extent when heated. They have low coefficients of linear expansion.

Mechanical properties

* **Strength** – This is the ability of a material to resist an applied force (load) without fracturing (breaking). It is also the ability of a material not to yield. Yielding is when the material 'gives' suddenly under load and changes shape permanently but does not break. This is what happens when metal is bent or folded to shape. The load or force can be applied in various ways. (You must be careful when interpreting strength data quoted for various materials. A material may appear to be strong when subjected to a static load, but break when subjected to an impact load. Materials also show different strength characteristics when the load is applied quickly rather than slowly.)

* **Toughness** – This is the ability of a material to resist impact loads. Strength and toughness are not the same. For example, hardened steel has great strength but lacks toughness.

* **Elasticity** – Materials that change shape when subjected to an applied force, but spring back to their original size and shape when that force is removed, are said to be elastic.

* **Plasticity** – Materials that flow to a new shape when subjected to an applied force and keep that shape when the applied force is removed are said to be plastic.

* **Ductility** – Materials that can change shape by plastic flow when they are subjected to a pulling (tensile) force are said to be ductile.

* **Malleability** – Materials that can change shape by plastic flow when they are subjected to a squeezing (compressive) force are said to be malleable.

* **Hardness** – Materials that can withstand scratching or indentation by an even harder object are said to be hard. Most metals, such as iron and steel, are hard.

* **Rigidity** – Materials that resist changing shape under load are said to be rigid. The opposite is flexibility. Rigid materials are usually less strong than flexible materials. For example, cast iron is more rigid than steel but steel is stronger and tougher. However, the rigidity of cast iron makes it a useful material for machine frames and beds. If such components were made from a more flexible material, the machine would lack accuracy and it would be deflected by the cutting forces.

Activity

State the properties required of materials used in each of these applications:

(a) A wire rope sling used for lifting heavy loads.

(b) The axle of a motor vehicle.

(c) A spring.

(d) A support for a cable carrying high-voltage electricity.

(e) The handle of a soldering iron.

(f) The fuselage of a light aircraft.

Just checking

* What are the main classes of material used in engineering and why are their properties important?
* What are the main chemical, electrical, magnetic, thermal and mechanical properties of materials?
* What is corrosion and what causes it?
* What is degradation and what causes it?
* What is the purpose of a protective surface coating and what types of coating are available?

Electrical components

Electrical and electronic components are assembled to form complete circuits. Individual components have different functions, such as switching a supply on and off, or reducing the current in a circuit to a safe value. As an engineer, you will need to get to know what these components do and what they look like.

Types of electrical component

A wide variety of electrical components is used in engineered products.

* **Cells and batteries** – A cell is a source of **direct current (d.c.)** electrical energy. Primary cells have a nominal potential of 1.5 V each. They cannot be recharged and are disposable. Secondary cells are rechargeable. Lead-acid cells have a nominal potential of 2 V, nickel cadmium (NiCd) and nickel metal hydride (NiMH) cells both have a nominal potential of 1.2 V. Cells are often connected in series to form a battery with increased overall voltage. For example, a 12 V car battery has six lead-acid secondary cells, each of 2 V.

* **Resistors** – Resistors control the magnitude of the current flowing in a circuit. The resistance value of a resistor may be fixed, pre-set by the user or variable in use. The electric current does work in flowing through a resistor and this heats it up. The resistor must be chosen so that it can withstand this heating effect and sited so that it has adequate ventilation.

* **Capacitors** – Capacitors may also be fixed, pre-set by the user or variable in use. Capacitors store electrical energy but, unlike secondary cells, may be charged or discharged almost instantaneously. The stored charge is much smaller than the charge stored by a cell or battery. Large-value capacitors are often used in power supplies while medium-value capacitors may couple signals in audio frequency amplifiers. Small-value capacitors are used in radio and TV receivers, where they are often found in the tuned (resonant) circuits that select or reject signals at a particular frequency.

* **Inductors** – Inductors act like electrical 'flywheels'. They limit the build up of **alternating current (a.c.)** in a circuit and try to keep the circuit running by putting energy back when the supply is turned off. They are used as current limiting devices in fluorescent lamps. Small-value inductors are used in radio and TV receivers, where they work with capacitors in tuned (resonant) circuits to select or reject signals at a particular frequency.

* **Switches** – Switches control current flow in a circuit. They only open or close, so the current either flows or it doesn't. Various types are available, including those that can open or close several circuits at the same time. A special type of lever-actuated switch, called a microswitch, is often used in conjunction with motorised systems and actuators, where it can be used to initiate or limit movement, or change direction.

* **Relays** – Relays are another form of switching device. They can be either electromechanical (using a coil and a set of contacts operated by electromagnetism) or solid-state (using a semiconductor switching device). Relays are capable of controlling high-voltage/high-current circuits from much lower voltage/current. For example, a relay capable of switching a 220 V a.c. supply at 10 A (a load of 2.2k W) can be controlled from as little as 0.1 A at 12 V. Contactors are similar to relays and are used for holding a supply when operated.

* **Fuses** – Fuses protect a circuit from excess current. This can result from a fault in the circuit, in an appliance connected to the circuit or through connecting too many appliances to one circuit. The current heats up the fuse wire. If the current reaches a pre-determined value, the fuse wire melts and breaks the circuit. Without a fuse, the wiring could overheat and might catch fire.

* **Circuit breakers** – Circuit breakers are also used to protect circuits. Unlike conventional fuses, they operate quickly and can be sensitive to small currents. Various types are available, including miniature circuit breakers (MCB) and residual current devices (RCD).

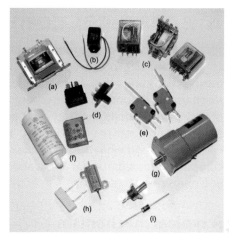

Some electrical and electronic components: (a) transformer, (b) sounder, (c) relays, (d) switches, (e) microswitches, (f) capacitors, (g) small d.c. motor, (h) resistors, (i) diodes

Direct current (d.c.) – current that flows in one direction. Conventional current flow is from positive to negative. Electrons flow in the opposite direction!

Alternating current (a.c.) – current that continuously changes direction, flowing first one way then the other. UK mains supply changes direction fifty times a second (a frequency of fifty cycles per second, or 50 Hz).

* **Transformers** – Transformers raise or lower the voltage of alternating currents (a.c.). Inductors and transformers cannot be used in d.c. circuits. The input side of a transformer is the 'primary'; the output side is the 'secondary'. If you increase the voltage, you decrease the current, so that (neglecting losses) the power ($V \times I$) is the same on both sides of the transformer.

* **Diodes** – Diodes act like non-return valves. They allow current to flow in one direction only and are used as rectifiers to change (or rectify) a.c. into d.c.

* **Transistors** – Transistors are used as amplifiers and as electronic switches (where they are capable of operating at high speed without moving parts).

* **Integrated circuits** – Integrated circuits contain all the components necessary to produce amplifiers, oscillators, central processor units, computer memories and a host of other devices fabricated onto a single slice of silicon, each chip being housed in a single compact package.

* **Sounders** – Sounders generate audible signals. They include electromechanical buzzers and piezoelectric transducers. Loudspeakers generate audible signals but are more suited to reproducing speech and music.

* **Motors** – Motors produce motion when electric current is fed to them. Many types are available, operating from d.c. and a.c. They are often fitted with some form of reduction drive to provide slow rotation speeds.

Selecting electronic components

Large-scale manufacturers of electronic equipment would buy their components directly from the makers in bulk. For small-scale batch production and prototype work, it is usual to buy from a wholesale or retail supplier on a 'one-stop' purchasing basis.

Activity

A small radio-manufacturing company has asked you to advise them on the selection of weatherproof loudspeaker for use in a portable 'sport' radio. The loudspeaker should have the following specification:

Impedance: 8 Ω (nominal)

Diameter: 70 mm (maximum)

Power rating: 0.5 to 1 W

Use electronic component suppliers' catalogues to locate a suitable loudspeaker drive unit and specify its part number or stock code and cost when purchased in multiples of 50 units.

Just checking

* What main types of electrical/electronic component are available?
* What is the difference between a.c. and d.c.?
* What must you consider when selecting an electrical/electronic component for an application?

Mechanical components

Just as electrical and electronic components are assembled to form complete circuits, mechanical components are assembled to form **mechanisms** and other more complex arrangements. Individual mechanical components have different functions, such as joining parts of an assembly, changing the line of action of a force or transmitting a torque. Engineers need to develop a good understanding of these components and what they look like. Once again, choosing the right component for a particular application is an important skill that all engineers need to develop.

A typical mechanism

A typical mechanism is shown in the photograph. This is used in a machine that drives a heavy roller alternately forwards and backwards. The mechanism uses a belt drive from an a.c. motor to a system of gears and, just like an electronic circuit, uses a number of individual components (some identified in the diagram).

Types of mechanical component

Many different mechanical components are used in engineered products:

* **Screwed fastenings** – These include nuts, bolts, screws, studs and washers. They come in a wide range of sizes and types of screw thread. There are also a variety of heads, usually a compromise between strength, appearance and ease of tightening. The hexagonal head is usually selected for general engineering applications. The more expensive cap-head screw is widely used in manufacture of machine tools, jigs and fixtures, and other highly stressed applications. These are forged from high-tensile alloy steel, thread-rolled and heat-treated. By recessing the cap-head, a flush surface is provided for safety and easy cleaning. Various washers are used with screwed fastenings, including locking and tab washers.

* **Riveted joints** – Rivets provide an alternative method of joining sheet metal and other thin components that have a flat profile. Rivets are made from malleable materials and are formed by hammering. Riveted joints are very strong providing they are correctly designed and assembled. The joint must be designed so that the rivet is in shear and not in tension. When selecting a rivet, you need to consider the material used for the rivet as well as the shape of its head.

* **Geared drives** – Gears provide a means of transmitting power. Types include straight tooth, helical, bevel, worm and rack and pinion drives. They are frequently used to change the speed at which a shaft rotates. For example, by using a 50:1 gear drive with a motor that revolves at 300 revolutions per minute, we can produce a drive shaft that revolves once every second. Gear trains can be simple or complex and may

Mechanism – arrangement of mechanical components that perform a function, such as opening a window

Fastening – a means of securely joining two parts, such as a panel and its supporting frame or a machined component to its mounting block

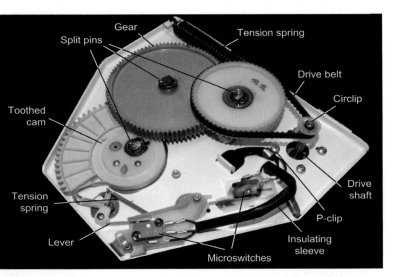

A typical mechanism showing some of the main components

involve additional idler gears that can be used to change the direction of rotation and/or increase the distance between the driver and the driven gears.

* **Belt and chain drives** – Belt and chain drives are an alternative method of transmitting power. They use driver and driven pulleys or sprocket wheels to transmit power, using a belt or chain respectively. This type of drive is often simpler than a geared drive and is preferred when there is an appreciable distance between the driver and driven shaft. Various types of drive belt are available. They must be resistant to fatigue failure, temperature change and a wide range of environmental conditions.

* **Shafts and keys** – Shafts and keys are used to transmit power in rotary systems. Keys can be straight, tapered or semi-circular (Woodruff keys) and they are used to lock the ends of the shaft to a gear or pulley. If excessive torque is applied to a shaft, the key will break and provide protection for the drive and/or driven components.

* **Bearings** – Bearings are used to support shafts and allow rotation with minimal friction. Ball bearings comprise a cage, in which the balls are guided, along with the raceway segments and integral seals. Bearings can be sealed, maintenance-free units or they may require periodic inspection and lubrication, using an appropriately rated grease.

Selecting mechanical components

When selecting mechanical components, you need to consider various factors.

* **Mechanical, chemical and thermal properties** – Is it strong enough for the application? Will it resist corrosion/degradation over the full range of service conditions? Will it wear out due to friction?

* **Compatibility** – Is the material of which the component is made compatible with the other materials used in the application? Is it softer or harder than other components? Will it suffer expansion or contraction at a different rate? Will it wear out at a different rate? If it fails, will it damage other components?

You may also need to consider other factors.

* **Fitting** – How is the component fitted (for example, screw threads, keys, etc.). Is the screw thread suitable for the job? Will it be necessary to use a locking device to counteract vibration? Note that coarse threads are stronger than fine threads, particularly in soft metals such as aluminium, but fine threads are less likely to work loose. Many types of locking device are available, including lock nuts, split pins and tab washers.

* **Physical dimensions** – For example, length and diameter. Will the component fit in the available space? Will it interfere with other components?

* **Access to the component** – For example, ease of inspection and/or replacement in the event of failure.

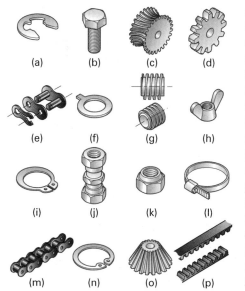

A selection of mechanical components: (a) e-clip (used for retaining a rod or a shaft) (b) hexagonal threaded bolt (c) helical gear (d) straight gear (e) chain link (f) tab washer (g) worm gear (h) wing nut (i) external circlip (used for retaining a rod or a shaft (j) nut coupler (k) anti-vibration nut (l) hose clip with integral worm drive (m) chain (n) internal circlip (o) bevel gear (p) toothed belt.

> ## Activity
>
> Identify the mechanical components shown in the photograph. Briefly explain the function of each component. Use a mechanical component supplier's catalogue to locate a similar component and state its catalogue reference or stock code.

> ## Activity
>
> Identify FOUR different types of bolt fixing. Illustrate your answer with a sketch and list three factors you would need to consider when selecting a bolt fastening for a particular application.

> ## Just checking
>
> * What main types of mechanical component are available?
> * What do you need to consider when selecting a mechanical component for a given application?

Marking out – the process of preparing a component for manufacture by applying surface marks that correspond to features of the finished component (for example, the position of holes

Surface finish – a coating or other finish applied to an engineered product to improve appearance, durability or corrosion resistance

Tools and processes

In order to manufacture an engineered product, or deliver an engineered service, you will need to use a variety of tools and processes. These differ widely according to the type of engineered product or engineering service. Obviously, the tools and processes used in the manufacture of a mobile phone are very different from those used in a motor vehicle repair workshop. As an engineer you must get to know how to use the tools and processes that are relevant to your chosen branch of engineering. You will also benefit from an understanding of those tools and processes that are more generally used.

Stages in the production of an engineered product

Making an engineered product includes the selection of appropriate:

* **materials and components** – for example, metals, composite materials.
* **processes** – for example, for measurement, shaping and finishing.
* **tools and equipment** – for example, pillar drills or soldering irons.
* **inspection and quality assurance measures.**

Measurement

Accurate measurement is essential when carrying out most engineering tasks. Before you can mark out a component or check it during manufacture, you need to know how to make an accurate measurement. All engineering measurements are comparative processes. This means that you need to compare the size of the feature to be measured with that of a known standard. Here are some tools used for measurement.

* **Steel rules** – The distance between marks, or the width of the work, is being compared with the rule. In this instance, the rule is our standard of length. It should be made from spring steel and the markings engraved into the surface of the rule. The edges should be ground so that it can be used as a straight edge, and the datum end of the rule should be protected from damage so that accuracy is not lost.

* **Callipers** – These are used to increase the usefulness of the steel rule and improve accuracy when taking measurements. They are used to transfer the distances between the faces of the work to the distances between the lines engraved on the rule. A steel rule can only be read to an accuracy of about ± 0.5 mm. This is rarely accurate enough for precision engineering purposes. To improve accuracy, callipers fitted with vernier or micrometer scales are used. Odd-leg (hermaphrodite) callipers are used as an aid to scribing lines when marking out (see later).

* **Try-squares and protractors** – Try-squares are used for measuring right angles, while protractors are used for measuring all other angles.

Marking out

Marking out is essential when manufacturing mechanical parts. It provides guidelines to work to. It also controls the size and shape of the work piece and the position of holes, slots or guides that need to be cut in it. Marking out usually involves scribing lines and centre marks.

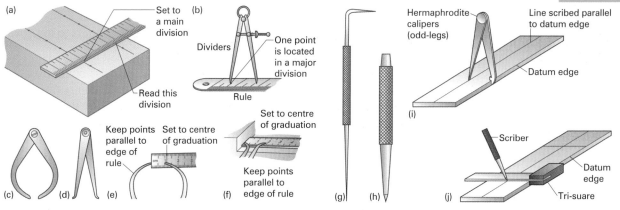

Tools and methods used for measurement

(a) using a steel rule (b) using dividers (c) outside callipers (d) inside callipers (e) using outside callipers (f) using inside callipers (g) scriber (h) centre punch (i) using odd-leg callipers (j) using a scriber and try square

Scribed lines are fine lines cut by the point of a scribing tool (scriber) into the surface of the material being marked out. To ensure that the lines show up clearly, the surface of the material is coated with a thin film of a contrasting colour. For example, the surfaces of a casting to be machined are often whitewashed. Bright metal surfaces can be treated with a marking out 'ink'. Plain carbon steels can be treated with copper sulfate solution, which copperplates the surface. This has the advantages of permanence but take care: copper sulfate will attack any marking out and measuring instruments with which comes into contact.

Centre marks are made with a dot or centre punch. A dot punch has a fine conical point with an included angle of about 60°. A centre punch is heavier and has a less acute point angle of about 90°. It is used for making a centre mark for locating the point of a twist drill, preventing the point from wandering at the start of a cut.

The dot punch is used for two purposes. Firstly, a scribed line can be protected by a series of centre marks made along the line. If the line is accidentally removed, it can be replaced by joining up the centre marks. Secondly, dot punch marks are used to prevent the centre point of dividers from slipping when scribing circles and arcs of circles. Note that, when a centre punch is driven into the work, distortion can occur. This can be a burr raised around the punch mark, swelling of the edge of a component, or the buckling of thin material.

The basic tools required for marking out are:

* **scriber** – to produce a line

* **steel rule** – to measure distances and act as a straight edge to guide the point of the scriber

* **dividers** – to scribe circles and arcs of circles (where necessary)

* **callipers and try-squares** – (note that odd-leg callipers are used to scribe lines parallel to a datum edge while a try-square and scriber are used to scribe a line at right-angles to a datum edge).

Activity

Use the production plan for the simple LED flasher that you produced in Topic 3 to manufacture the LED flasher. Apply a suitable finish and labelling to the product (you may need to modify your production plan to take this into account) and ensure that you keep a full record of each stage of production (a digital camera might be useful here!).

Surface finish

Surface finishes can be applied to many engineered products to improve appearance, durability or corrosion resistance.

* **Grinding** – A grinding wheel consists of abrasive particles bonded together. It does not 'rub' the metal away – it cuts like any other cutting tool, as each particle is a cutting tooth. It is like a milling cutter with thousands of teeth. Wheels are made in many shapes and sizes. They are also available with a variety of abrasive particles and bonding materials. It is essential to choose the correct wheel for any given job.

* **Polishing** – This produces a better finish than grinding but only removes the smallest amounts of metal. It gives a smooth and shiny surface but the geometry of the surface is uncontrolled. Polishing is used to produce decorative finishes, to improve fluid flow (e.g. through the manifolds of racing engines) and to remove machining marks from surfaces that cannot be precision ground. This reduces the risk of fatigue failure in highly stressed components.

* **Electroplating** – This coats components with another metal that is more decorative and/or corrosion resistant.

* **Hot-dip galvanizing** – Hot-dip galvanizing coats low carbon steels with zinc without using an electroplating process. It is the original process used for zinc-coating buckets, animal feeding troughs and other farming accessories. It is also used for galvanized sheeting. The work to be coated is chemically cleaned, fluxed and dipped into the molten zinc. This forms a coating on the work. A small percentage of aluminium is added to the zinc to give a bright finish. The molten zinc also seals any cut edges and joints in the work and renders them fluid tight. Metal components may also be coated with non-metallic surfaces.

* **Oxidising (blueing)** – Steel components have a natural oxide film due to reaction with atmospheric oxygen. This can be thickened and enhanced by heating the component until it takes on a dark blue colour; then immediately dipping it into oil to seal the oxide film. The process does not work if there is any residual mill scale on the metal surfaces. Alternatively, an even more corrosion resistant oxide film can be applied to steel components by chemical blacking. The components are cleaned and degreased. They are then immersed in an oxidising chemical solution until the required film thickness has been achieved. Finally the treated components are rinsed, dried and oiled. Again, the process only works on bright surfaces.

* **Plastic coating** – Plastic coatings can be functional, corrosion-resistant and decorative. There is a wide range of materials available in a variety of colours and finishes. They can give electrical and thermal insulation, abrasion resistance, cushioning effects with coatings up to 6 mm thick and non-stick properties (Teflon PTFE coatings). They can also give permanent protection against weathering and atmospheric pollution, resulting in reduced maintenance costs, resistance to corrosion by a wide range of chemicals, covering of welds and the sealing of porous castings.

* **Fluidised bed dipping** – Finely powdered plastic particles are suspended in a current of air in a fluidising bath. The powder continually bubbles up and falls back, looking as though it is boiling. The work is preheated and immersed in the powder. A layer melts onto the surface of the metal to form a homogeneous layer.

* **Liquid plastisol dipping** – This process is limited to PVC coating. Plastisols are a form of PVC that is liquid at normal room temperatures.

Activity

Reading vernier calipers

What is the indication on the vernier calliper shown in the photograph? Explain how you arrive at your answer.

Process	Applicable materials	Available tools	Typical applications
Adhesive bonding	Most materials but particularly suitable for plastics and composite materials	Applicators for resin-based adhesives; hot and cold melt glue guns	
Bending	Soft metals such as aluminium, copper, tinplate		Bending sheet metal; bending tubes and pipes
Crimping		Crimping tool	Terminating electrical cables and wires with suitable connectors
Cutting	Most metals, plastics and composite materials	Hacksaw; bandsaw; file (various types); guillotine (sheet metal); metal shears; wire cutters; pipe cutters; bolt cutters	
Drilling	Cutting holes in sheet metal; cutting 'blind' holes in cast metal		Cutting holes in sheet metal; cutting 'blind' holes in cast metal
Grinding		Straight grinder, angle grinder, off-hand grinder	Removing surface imperfections in metal parts; removing excess metal from castings
Heat treatment (hardening)	Carbon steels		Hardening steel components such as gears and other parts used in power transmission
Machining		Lathe (various types), milling machine (various types)	Surface facing (producing flat surfaces); cutting slots in cast metal
Soldering	Copper and brass components; copper wires	Soldering iron; blowlamp	
Welding	Steel in various forms (sheet, plate, bar, angle, beam, channel, etc.)		Joining plates; making structural joints; joining decorative parts

Components are dipped in plastisols to form a coating and then heated to 'cure' the material and achieve the desired physical properties. No dangerous solvents are present during this process.

✳ **Painting** – Painting is used to provide a decorative and corrosion-resistant coating for metal and other surfaces. It is the easiest and cheapest means of coating that can be applied with any degree of permanence. It may be applied by brush, spray or dipping. It has three components: pigment (fine powder that gives opacity and colour), vehicle (a film-forming liquid or binder in a volatile solvent) and solvent/thinner (controls the consistency of the paint and its application). The binder is a natural or synthetic resinous material. When dry (set), it must be flexible, adhere strongly to the surface being painted, corrosion-resistant and durable. The solvent/thinner forms no part of the final paint film, as it evaporates. As it does so, it increases the concentration of catalyst in the 'vehicle' causing it to change chemically and set.

Activity

Copy the 'Processes' table onto an A4 sheet and use library and Internet resources to complete the missing boxes.

Just checking

✳ What equipment is used for measuring parts and components?

✳ What is marking out and why is it important?

✳ What types of surface finishes are available and why are they important?

Inspection

Inspection is an integral task in the production of an engineered product or delivery of an engineered service. A properly designed inspection process provides a means of checking that the various stages of production are working properly, that the materials and components are fit for purpose, and that the finished product or service will conform to the original design specification. Inspection is also important in the context of engineering maintenance. Few of us would be willing to fly in an aircraft that had not been regularly inspected!

The need for inspection

During the manufacture of a product, it is necessary to ensure that the various parts and components conform fully to specification. Not only does this provide a check on the production process and the materials/ components used; it also acts as a means of ensuring the product conforms to the design specification.

Routine inspection is designed to ensure that a product, plant or equipment conforms to specification and is operating correctly. It can also be useful in preventing breakdown, so is considered part of the routine maintenance of many engineered products.

Many products benefit from routine maintenance in the form of inspection and adjustment. Inspection may identify parts or consumables (such as grease or oil) that may need replacement. Inspection may not just be visual – it can also involve smell, touch and sound!

Mechanical inspection

Mechanical inspection can involve checking the size and 'fit' of components. When part of routine maintenance, it can also detect wear and early signs of mechanical failure. Various techniques are used, including tolerancing and gauging.

Nominal size	100 mm
Limits (low)	99.8 mm
Limits (high)	100.2 mm
Tolerance	0.4 mm
Deviation	± 0.2 mm
Mean size	100.0 mm

Tolerancing and gauging

Since no product can be made to an exact size, nor measured exactly, the designer often specifies dimensions by quoting an upper and lower size. If the component lies anywhere between these, it will be within specification and therefore function correctly. The closer the limits, the more accurately the component will work, but the more expensive it will be to make. A major advantage of using tolerance dimensions is that they can be checked with gauges without having to be measured. This is easier, quicker and much cheaper. Various types of gauge are commonly used, including:

* **plug gauges** – to check hole sizes
* **radius gauges** – to check the rounding of corners
* **feeler gauges** – to check the gap between parts

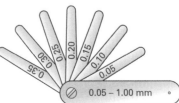

A feeler gauge

Check it out!

Imagine that you have been given £2,000 to purchase a second-hand car. You have identified the make and model of the car that you want to buy and have a shortlist of three vehicles to inspect. Make a list of at least ten items that you would wish to inspect, with reasons why you have included them. Compare your list with those produced by two or three other students. Did you pick the same ten items?

Personal learning and thinking skills

Investigate the use of a feeler gauge to check the gap in a spark plug. Produce an A3 illustrated poster for a workshop showing how this tool should be used and stored when not in use.

Inspection – examination by sight, touch, smell and sound (as appropriate) to determine the functional state of a product, process plant or equipment

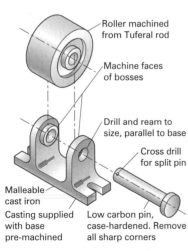

Medicine component

Roller machined from Tuferal rod

Machine faces of bosses

Drill and ream to size, parallel to base

Cross drill for split pin

Malleable cast iron

Casting supplied with base pre-machined

Low carbon pin, case-hardened. Remove all sharp corners

* **thread gauges** – to check the pitch of screw threads
* **calliper gauges** – to check the thickness of components.

A jet aircraft undergoing detailed inspection during a major service check

Activity

Prepare an inspection plan for the LED flasher that you produced in Topic 7. The design specification is given in the table. Carry out appropriate tests and measurements to ensure that the LED flasher is operating to this specification. If necessary, carry out a calibration adjustment of the unit. Ensure that you keep a full record of each stage, a digital camera might be useful.

Specification	Value and tolerance
Supply voltage	9 V (nominal) with operation down to 6 V
Supply current	8 mA (± 2mA) at 9 V
Flash rate	1 s (± 10%) at 9 V

Activity

Prepare an inspection plan to assess the quality of the mechanical component shown on the previous page. Photocopy and enlarge the view of the component then mark the inspection points on it. Your inspection procedure should refer to critical features (including dimensions).

Electrical inspection

Electrical inspection usually involves measuring electrical specifications, such as output voltage or output current. It can also involve measuring the insulation resistance of a component. This is the resistance between a 'live' conductor and parts of the equipment that are normally earthed or 'grounded'. Insulation resistance is important because any significant leakage of current to earth can be unsafe, or may cause an earth leakage circuit breaker (ELCB) or residual current device (RCD) to trip.

Equipment used for electrical inspection includes:

* **voltmeters** – to check voltages
* **ammeters** – to check current
* **ohmmeters** – to check resistance and 'continuity'
* **insulation testers** – sometimes referred to as 'meggers', to check insulation resistance.

In addition, specialised test equipment, such as multimeters, oscilloscopes (see Unit 5 Topic 11) and signal generators, may be used to inspect and carry out detailed checks on electronic equipment.

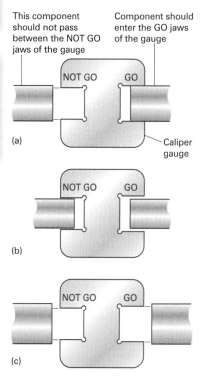

Use of a calliper gauge to check the thickness of a component using a 'go' and 'not go' method: (a) a correctly sized component enters the 'go' jaws but not the 'no go' jaws (b) an undersize component enters both the 'go' and the 'not go' jaws (c) an oversize component

Just checking

* What is inspection and why is it needed?
* What is a maintenance procedure and what should be included in it?

Unit 4 Assessment Guide

In this unit, you will investigate how to plan and manufacture a product. This involves selecting materials, components, tools and equipment and carrying out a risk assessment to ensure safe working practices are in place. You will develop the skills needed to achieve a logical and structured approach to problem solving.

Time Management

Manage your time well as this unit has a number of different components that will have to be researched. Ensure that you keep your work safe and that any work in electronic format has a secure and safe backup.

Be well organised. This is your chance to show that you are an independent enquirer, creative thinker, reflective learner, self manager and effective participator. These will contribute towards achievement of your Personal learning and thinking skills.

Plan ahead for your work experience, making a list of things that you need to find out or observe so that you make the most efficient use of your time.

Useful Links

Make good use of your work experience to find out as much as possible about issues that are relevant to your coursework. You should work and meet with people involved with production planning, manufacturing, quality control and health and safety. Talk to them about how products are manufactured and make a note of what they say.

There are a number of websites to help you with assessment focus 4.1, but remember to search UK sites only.

There is a wide range of manufacturers' data (web-based and hard copy catalogues) to help you with assessment focus 4.3.

Things you might need

Your evidence for assessment needs to be presented as an A4 process portfolio and should be in e-format. Your teacher should give you access to the required software to enable the correct presentation.

A digital camera or mobile phone would be useful, so that you record evidence of practical activities.

Remember that you will have to carry out workshop activities in order to complete assessment foci 4.1, 4.4 and 4.5. Make sure you obtain witness statements from your supervisor and photographs showing you working correctly and safely.

Remember to maintain a focus on planning and safe working throughout your project and the need to use a variety of skills which are linked.

How you will be assessed

What you must show that you know	Guidance	To gain higher marks
That there are health and safety procedures which must be followed when working in areas identified as hazardous through risk assessment. *Assessment focus 4.1*	✳ Identify Health and Safety procedures which apply to the practical work you are going to carry out. ✳ Identify key safety points when working in a workshop. ✳ Describe the use of personal protective equipment (PPE).	✳ You need to describe and compare the responsibilities which you and others have, when using tools and equipment in a workshop. ✳ You need to risk assess activities which you have identified as hazardous. ✳ Obtain witness statements and take photographs to prove that you carried out risk assessment.
How to plan the manufacture of an engineered product or the carrying out of a service. *Assessment focus 4.2*	✳ Choose either a product to manufacture or a service to carry out. ✳ Agree the choice with your teacher. ✳ Produce a production plan which has information about: 　✳ how you intend to make the product or carry out the service. 　✳ materials and components 　✳ tools and equipment 　✳ sequence of events	✳ You need to explain the reasons behind your choice of tools and equipment. ✳ You should review how successful you were at manufacturing the product or carrying out the service (assessment foci 4.4 and 4.5). ✳ You must then propose ways to improve your manufacturing or servicing methods and incorporate these into a revised production plan.
How to select the materials, parts and components to be used when manufacturing a product or carrying out a service. *Assessment focus 4.3*	✳ Using your production plan identify and select the materials, parts and components you will need to make the product or carry out the service ✳ Prepare the materials, parts and components for use e.g. electric cable from a drum being cut into short manageable lengths.	✳ You need to explain why you chose the materials, parts and components. ✳ You need to justify why you prepared the materials, parts and components in the ways that you did.
That you can use tools and equipment to manufacture a product or carry out a service. *Assessment focus 4.4*	✳ Manufacture the product or carry out the service with guidance from your teacher. ✳ Obtain witness statements and take photographs to prove that you carried out the practical work effectively.	✳ You need to carry out the practical work without any help. ✳ You must have used additional drawings or documentation to support your production plan. ✳ You must have worked safely and not placed yourself or anyone else in jeopardy.
That you can use equipment and inspection procedures to quality check a product which you have manufactured or a service which you have carried out. *Assessment focus 4.5*	✳ Perform inspection checks on the product you manufactured or the service which you carried out. ✳ Obtain witness statements and take photographs to prove that you carried out the practical work effectively. ✳ Present the results of your inspection checks as a written statement.	✳ You should record the key measurements taken during inspection. ✳ You should present the results of inspection as a report which confirms whether the product or service is to specification or not. ✳ You should review the production plan you produced for assessment focus 4.2.

5

Introduction

Electronics has a huge impact on the way we live. We listen to the radio, watch TV, talk on our mobile phones, and play on computers and games consoles. Electronic management systems are used on cars, aircraft and ships; satellite-based systems, and other electronic aids, such as radar and navigation systems. Electronic signalling is used on our roads and railways, and electronic surveillance helps to ensure our security and protect our property both while we are on the move and when we are at home. All of this serves to underpin the value and importance of electronics both in the home, at work and in our leisure activities!

This unit will provide you with an insight into the exciting and challenging world of electronics. It will help you understand how basic electrical/electronic circuits work and how to work safely with them. The unit will introduce you to a variety of common electronic components, and how to recognise and select them. The unit will also provide you with an opportunity to construct and test a prototype electronic circuit, and show you how to locate and rectify simple faults.

This unit is assessed by your tutor. On completing it you should:

1. Understand safe working practices in the workshop/laboratory and understand relevant electrical and electronic principles.

2. Be able to recognise and select components used in electrical and electronic circuits.

3. Be able to construct an electronic circuit and understand its basic operating principles.

4. Be able to test and fault find on electronic circuits.

Thinking points

Think about the following key points as you work through this unit:

1. What hazards exist in electrical/electronic circuits and how can they be avoided?

2. What are the basic principles on which electrical/electronic circuits operate?

3. What are the basic electronic components and how are they selected?

4. What equipment and techniques are used for constructing electronic circuits?

Electricity

In today's world, electricity is something that we all take for granted. Just think, for a moment, about what electricity means to you and how it affects your life. To be able to understand it, you will need to have a grasp of some of the basic principles that underpin electricity and electronics. These include fundamental concepts such as charge, current, resistance and Ohm's Law.

Our invisible friend

How many of the things that you use every day depend on electric current? But of course you can't actually see it! You only know that it's there by what it does, giving you heat, light, motion and sound.

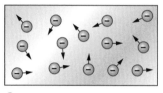

⊖ Free electron
Metal conductor
(no current flowing)

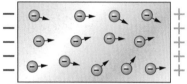

Metal conductor
(current flowing)

Current – the organised movement of charge carriers (electrons in metal conductors)

Conductor – a material that conducts electric current. Metals are good conductors

Insulator – a material that does not conduct electric current. Glass, ceramics and plastics are good insulators

Resistance – the opposition to flow of current

Potential difference – the voltage drop that appears across a circuit or component when an electromotive force (e.g. a battery) is connected across it

Conductors and insulators

The materials that we use in engineering are often classed as either electrical **conductors** or electrical **insulators**. Conductors are materials that have many free electrons, each of which carries a tiny amount of negative electric charge. Conductors include metals like aluminium, copper, gold and iron. Insulators, on the other hand, have no free electrons available to carry electric charge. Insulators include glass, ceramic, and plastic materials.

Electric current

When there is no external electric field present, the motion of free electrons in a conductor is random and the electrons simply drift around. However, if an external electric field is applied to a conductor by connecting a battery or other source of electromotive force (e.m.f.) to it, the motion of the electrons changes so that the negatively charged electrons drift towards the positive end of the conductor. In a metal conductor, electric current is simply the organised movement of electrons.

Current is defined as the rate of flow of electric charge and its unit is the Ampere, A. One Ampere is equal to one Coulomb of charge transferred in one second. Writing this as a formula gives:

$$I = \frac{Q}{t}$$

where I is the current in Amperes (A), Q is the charge in Coulombs (C), and t is the time in seconds (s).

Example

A steady current of 3 A flows for two minutes (i.e. 120 seconds). By re-arranging the formula, the amount of charge transferred will be:

$$Q = I \times t = 3 \times 120 = 360 \text{ C}.$$

Because of their negative charge, electrons will flow from a negative potential to a more positive potential (like charges attract and unlike charges repel). However, when we indicate the direction of current in a circuit, we show it as moving from a point that has the greatest positive potential to a point that has the most negative potential. We call this conventional current and, although it may seem odd, you just need to remember that it flows in the opposite direction to the movement of electrons!

Resistance

All materials at normal temperatures oppose the movement of electric charge. This opposition to the flow of the charge carriers is known as resistance and it results from collisions between the charge carriers (i.e. the electrons) and the atoms of which the material is composed. The symbol used for resistance is R and its unit is the Ohm which is often shown by the Greek symbol, Ω (omega).

Ohm's Law

The larger the resistance, the greater the opposition to current flow. If the temperature does not vary, the ratio of **potential difference** (i.e. voltage drop) across the ends of a conductor to the current flowing in the conductor is a constant. This relationship is known as Ohm's Law and it leads to the relationship:

$$\frac{V}{I} = \text{a constant} = R$$

where V is the potential difference (or voltage drop) in volts (V), I is the current in amps (A), and R is the resistance in ohms (Ω).

The formula may be arranged to make V, I or R the subject, as follows:

$$V = I \times R \qquad I = \frac{V}{R} \quad \text{and} \quad R = \frac{V}{I}$$

Example

A voltage drop of 6 V appears across a resistor of 120 Ω. The current flowing in the resistor will be:

$$I = \frac{V}{R} = \frac{6}{120} = 0.05 \text{ A (or 50 mA)}$$

Activity

Carry out an experiment to verify Ohm's Law. Take corresponding readings of voltage and current and use these to plot a graph. Use measured values of voltage and current to calculate the value of several unknown resistors.

Activity

See if you can answer these questions:

1 A charge of 180 C is transferred at a steady rate in a time of 5 minutes. What current flows in the circuit?

2 What voltage drop (potential difference) will be developed across a 56 Ω resistor if a current of 0.1 A flows in it?

3 What current will flow in an 18 Ω resistor when it is connected to a 9 V battery?

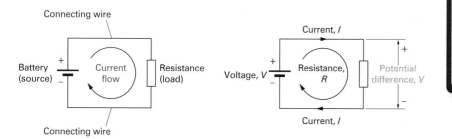

Voltage, current and resistance in a simple cicuit

Just checking

  ✳ What is an electrical insulator? Give three examples.
  ✳ What is an electrical conductor? Give three examples.
  ✳ What is electric current and what causes it to flow?
  ✳ What is the relationship between voltage, current and resistance?

Safe working practice

When working on electrical and electronic equipment, your safety and the safety of those around you should be paramount in everything you do. Electricity acts very quickly and can kill. Always think carefully before working on circuits where mains or high voltages (i.e. those over 50 V, or so) might be present. If you don't take care, there is a very real risk of electric shock.

Hazards and electric shock

Voltages in many items of electronic equipment, including all those using an alternating current (AC) mains supply, can cause sufficient current flow in the body to disrupt the heart. Bodily contact with mains or high-voltage circuits can be lethal. The most dangerous path for electric current within the body (i.e. the one that is most likely to stop the heart) is the one that goes from one hand to the other. The hand-to-foot path is also dangerous, but less so than the hand-to-hand path.

So, before you start to work on any electronic equipment, not only switch off but also disconnect the equipment at the mains by removing the mains plug. If you have to make measurements or adjustments on 'live' equipment, use one hand only. Place your 'spare' hand safely away from contact with anything metal, including the chassis of the equipment.

Fuses

A fuse is an electrical conductor designed to fail when the current passing through is too high. Most fuses comprise a thin wire conductor fixed inside a glass or ceramic tube, with caps at each end that connect to the fuse-holder. Fuses are usually designed to protect a battery or an a.c. mains supply against the effects of catastrophic failure. They can help protect the equipment from over-heating and damage. Always follow the manufacturer's recommendations when selecting and fitting fuses.

Application	Power rating	Fuse
Soldering iron	25 W	1 A
Personal computer	350 W	3 A
Halogen lighting	500 W	5 A
Electric fire	2 kW	13 A

Types of fuse fitted to electrical and electronic equipment, include:

* fuses designed for fitting into mains plugs (i.e. the 13 A domestic plug found in the UK, which may be fitted with 1 A, 3 A, 5 A and 13 A cartridge fuses)
* quick-blow fuses
* slow-blow (or time-delay) fuses.

Quick-blow fuses are usually designed to break within about one second. Time-delay fuses are designed to withstand currents well in excess of their rated current values for short periods (typically 5 to 10s). They are used where there may be a sudden surge of current on 'switch-on'.

Energy and power

Electrical energy is converted into other forms of energy by components such as resistors producing heat, loudspeakers producing sound, lamps and light emitting diodes producing light.

The unit of energy is the Joule (J). Power is measured in Watts (W). A power of 1 W is equivalent to energy being used at the rate of 1 J per second. We can find the power in a circuit from the relationship:

$$P = IV$$

where P is the power in Watts (W), I is the current in Amperes (A) and V is the voltage in Volts (V).

The formula may be arranged to make P, I or V the subject, as follows:

$$P = I \times V \quad I = \frac{P}{V} \quad \text{and} \quad V = \frac{P}{I}$$

Example

A current of 1.5 A is drawn from a 12 V supply. What power is supplied?

The power can be calculated from: $P = IV = 1.5 \times 12 = 18$ W

Example

A lamp is rated at 6 V, 0.3 W. What current does it use?

By re-arranging the formula to make I the subject we get:
$I = P / V = 0.3 / 6 = 0.05$ A (or 50 mA)

Personal learning and thinking skills

Use Internet and library resources to produce an A3 poster describing typical electrical hazards when working on electrical and electronic circuits. Include the procedure for dealing with electric shock.

Activity

Visit your workshop or laboratory and see if you can locate the electrical circuit breaker and residual current device. Make a sketch showing where these devices are located and explain how and when they are used.

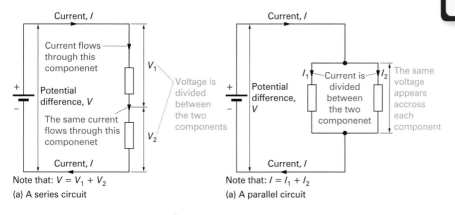

Note that: $V = V_1 + V_2$
(a) A series circuit

Note that: $I = I_1 + I_2$
(a) A parallel circuit

Circuit breakers and protection devices

In the UK, mains power in homes and offices is distributed by two main methods: a ring-main circuit, distributing power to 13 A sockets, or a radial circuit for lighting. There can be several independent circuits, each protected by a miniature circuit breaker (MCB) at the consumer unit.

If you touch a live conductor and are earthed through your other hand or your feet, you will receive an electric shock. The possibly dangerous consequences can be avoided by fitting a residual current device (RCD) or residual current circuit breaker (RCCB). Typically these operate at currents of 30 mA or 100 mA (30 mA giving greater protection).

Just checking

* What is alternating current and why are a.c. mains supplies dangerous?
* What does a circuit breaker do? How does an MCB differ from an RCD/RCCB?
* What is power and how does it relate to energy?
* What is the relationship between power, voltage and current?

Gaining control

In electronic circuits we need a means of controlling the flow of electric current. Resistors allow you to do just that. Look back at the previous topic and Ohm's Law. What else do resistors let us control?

Activity

Carry out an experiment on a resistor. Measure the current that flows in the resistor when different voltages are applied to it. Record your results in a table and plot a graph showing how the current varies with the voltage. Is this what you would expect? Use Ohm's Law to check your measured values.

Resistors

Resistors are used in almost every electronic circuit. They are used for controlling the flow of current and providing precise voltages. You need to know how to recognise them and be able to determine their resistance from their markings.

Resistor specifications

The specifications for a resistor usually include:

* the value of resistance (expressed in Ω, kΩ or MΩ)

* the accuracy or tolerance in relation to the marked value (quoted as the maximum permissible percentage deviation from the marked value)

* the power rating (which must be equal to, or greater than, the maximum expected power dissipation).

Temperature coefficient and stability are also important considerations in certain applications.

Resistance value

Resistors are available in standard decade (×10) ranges with a large number of different values (see Activity). The values available for common types of fixed resistor can range from less than 1 Ω to as much as 10 MΩ (10,000,000 Ω).

Tolerance

Some minor variation in resistance value is inevitable due to manufacturing tolerance. Thus the value marked on the body of a resistor is not its exact resistance. For example, a resistor marked 100 Ω and produced within a tolerance of ±10% will have a value between 90 Ω and 110 Ω . If a particular circuit requires a resistance of, for example, 105 Ω, a ±10% tolerance resistor of 100 Ω will be perfectly adequate. If, however, you need a component with a value of 101 Ω, then it would be necessary to obtain a 100 Ω resistor with a tolerance of ±1%.

Power rating

The power rating (or 'wattage rating') of a resistor is the maximum power that it can safely dissipate. Power ratings are related to operating temperatures and resistors should be derated (so that they operate with less power) at high temperatures.

Fixed resistor

Pre-set resistor

Variable resistor

Pre-set potentiometer

Variable potentiometer

Temperature sensitive resistor (thermistor)

Light dependent resistor (LDR)

Resistor markings

Carbon and metal oxide resistors are normally marked with **colour codes** that indicate their value and tolerance. Two methods of colour coding are in common use: one involves four coloured bands, while the other uses five. Other types of resistor have values marked on their bodies.

Types of resistor

Some common types of resistor are shown in the photograph below. These are (left to right):

* carbon film resistors with power ratings between 0.2 W and 2 W
* ceramic clad wirewound resistor rated at 7 W
* metal clad resistor rated at 15 W
* miniature skeleton preset **potentiometers**
* multi-turn preset potentiometer
* carbon track variable potentiometer
* multi-turn wirewound variable potentiometer
* light dependent resistor (LDR)
* temperature sensitive resistor (thermistor).

Example
First digit (brown) = 1
Second digit (black) = 0
Multipler (red) = 100
Tolerance (gold) = ± 5%
Value 10 x 100 = 1000 Ω = 1 kΩ, ±5%

Brown = 1   Gold = ±5%

Black = 0   Red = 2

1st and 2nd colour bands		3rd coloured band		4rd coloured band (tolerance)	
Black	0		Multiply by	Red	± 2%
Brown	1	Silver	0.01	Gold	± 5%
Red	2	Gold	0.1	Silver	± 10%
Orange	3	Black	1	None	± 20%
Yellow	4	Brown	10		
Green	5	Red	100		
Blue	6	Orange	1000		
Violet	7	Yellow	10000		
Grey	8	Green	100000		
White	9	Blue	1000000		

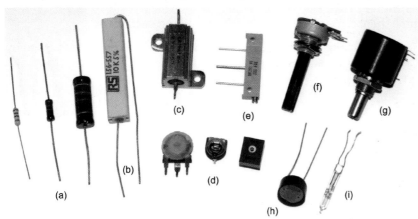

Some common types of resistor

We usually want the value of a resistor to remain the same but there are some types of component where the resistance may widely according to the temperature (a thermistor) or light (an LDR). These components can be useful in applications where we need to sense what's going on in the real world and use this as an input to an electronic circuit. Thermistors and LDR are often referred to as sensors.

Resistor – a component that introduces resistance into a circuit causing current flow to be reduced

Tolerance – the accuracy of a component, usually expressed as a percentage

Colour code – a system of coloured bands to indicate the value and tolerance of a component

Potentiometer – a variable resistor with three terminals, two with fixed resistance and one (the slider) that is varied between the other two

Capacitors

Getting charged up

Capacitors let us store a charge until we need it. The rate at which a capacitor can store (or release) its charge depends on the circuit components. How could we use this effect in simple timing applications?

Like resistors, capacitors are found in a wide range of electronic circuits. They are used for storing electric charge and then releasing it when needed. You need to know how to recognise them and to determine their capacitance from their markings.

Charge and discharge

A capacitor is, in effect, a reservoir into which charge can be deposited and then later extracted. It consists of nothing more than two parallel metal plates separated by air or an insulating **dielectric** material, such as polyester or polystyrene.

When a capacitor is being charged from a battery through a resistor, electrons will be attracted from the positive plate of the capacitor to the positive terminal of the battery. A similar number of electrons will move from the negative terminal to the negative plate. This sudden movement of electrons creates a brief surge of current, which then gets smaller and smaller until enough electrons have moved to make the e.m.f. between the plates the same as the battery. The capacitor is then said to be fully charged and an electric field will be present in the space between the two plates.

If we disconnect the battery, the positive plate is left with too few electrons, and the negative plate with a surplus. Since there is no path for current to flow between the plates, the capacitor remains charged with a potential difference between the plates. The capacitor can be discharged by simply placing a resistor across its terminals to drain away the charge until there is the same number of electrons on each plate.

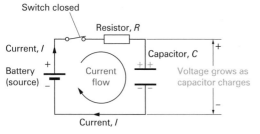

(a) Capacitor being charged

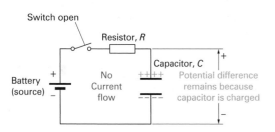

(b) Capacitor fully charged

The time constant, τ, of a capacitor-resistor circuit is given by the relationship:

$$\tau = C \times R$$

where C is the value of capacitance in Farads (F) and R is the resistance in Ohms (Ω).

The time taken to fully charge or discharge a capacitor is usually taken to be five times the time constant, or $5CR$.

Capacitance

The unit of capacitance is the Farad (F). A capacitor has a capacitance of 1 F if a current of 1 A flows in it when a voltage changing at the rate of 1 V/s is applied to it.

The charge or quantity of electricity that can be stored in the electric field between the capacitor plates is proportional to the applied voltage and the capacitance of the capacitor. Thus:

$$Q = CV$$

where Q is the charge in Coulombs (C), C the capacitance in Farads (F), and V the potential difference in Volts (V).

Example

A 100 µF capacitor is charged to a voltage of 25 V. The amount of charge stored will be:

$$Q = CV = 100 \times 10^{-6} \times 25 = 0.0025 \text{ C (or 2.5 mC)}$$

Capacitor specifications

The specifications for a capacitor usually include:

* the value of capacitance (expressed in µF, nF or pF)

* the accuracy or tolerance in relation to the marked value (quoted as the maximum permissible percentage deviation from the marked value)

* the voltage rating (which must be equal to, or greater than, the maximum expected applied voltage).

Voltage rating

Voltage rating is the maximum voltage that should be applied to the capacitor. Depending on the type, voltage ratings are quoted for direct current (d.c.) or alternating current (a.c.), with the a.c. voltage rating lower than the d.c.. Note that **electrolytic capacitors** need a d.c. voltage. The **polarising voltages** used for these can range from 1 V to several hundred volts, depending on the working voltage rating for the component. This voltage has to be applied with the correct polarity.

Capacitor markings

Most capacitors have written markings. The usual method for resin dipped polyester (and other) types quotes the value (µF, nF or pF), the tolerance (often either 10% or 20%) and the working voltage (often using _ and ~ to indicate d.c. and a.c respectively).

A three-digit code is also used in which the first two digits correspond to the first two digits of the value, while the third is a multiplier that gives the number of zeros to be added to give the value in pF. Other capacitors may use a colour code similar to that used for resistor values.

Activity

Carry out an experiment on a large value electrolytic capacitor. Measure the voltage that appears across the terminals at regular intervals while it is being charged from a d.c. supply. Record your results and plot a graph showing how voltage varies with time.

Charge – a negative charge results from an accumulation of electrons; a positive charge results from a lack of electrons.

Dielectric – an insulating material that separates the plates of a capacitor

Electrolytic capacitor – a capacitor that relies on chemical action for its operation

Polarising voltage – a d.c. voltage that needs to be applied to an electrolytic capacitor for it to operate

Fixed capacitor   Electrolytic capacitor

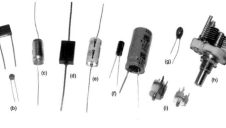

pre-set capacitor   Variable capacitor

Some common types of capacitor
a) Radial lead polyester b) Miniature ceramic disk
c) Tubular polystyrene d) Mica plate e) Axial lead electrolytic f) Radial lead electrolytic g) Tantalum bead electrolytic h) Air-spaced variable i) Solid dielectric preset

Just checking

* What is a capacitor and what is capacitance?
* What is the relationship between charge, voltage and capacitance?
* What happens when a capacitor is being charged and discharged?
* What types of capacitor are available?
* What symbols are used for showing capacitors on circuit diagrams?

Inductors and transformers

Electromagnetism is used in a number of useful components, including inductors, transformers, motors, generators, loudspeakers and relays. All of these components rely on the fact that a magnetic field is created whenever a current flows in a wire.

Inductance

Inductance is the property of a coil which opposes a change in the value of current flowing in it. Any change in the current applied to a coil/inductor will result in a voltage appearing across it. The unit of inductance is the Henry (H) and a coil is said to have an inductance of 1 H if a voltage of 1 V is induced across it when a current changing at the rate of 1 A/s is flowing in it.

Example

A current increases at a uniform rate from zero to 6 A in 2 s. If this current is applied to an inductor of 4 H, the voltage induced across the ends of the coil will be given by:

$$V = L \times \frac{I}{t} = 4 \times \frac{6}{2} = 12\,\text{V}$$

where V is the energy (in Joules), L is the capacitance (in Henries), and I is the current flowing in the inductor (in Amps).

Magnetic circuits

Batteries, resistors and capacitors can be arranged into circuits that show the links between the components. We do the same to help us understand magnetic components. In an electric circuit, current flows through the wires (conductors) that link the components together. In a magnetic circuit, it is the magnetic flux that links the components. Take a look at the diagram to see the difference between electric and magnetic circuits.

Transformers

Transformers consist of two coils wound on a common magnetic core. The input coil is called the primary winding while the output coil is called the secondary winding. Alternating current power can be coupled from one coil to the other. A particular advantage is that a voltage can be stepped up (a step-up transformer) or stepped down (a step-down transformer). This allows us to easily change an a.c. voltage. However, since no increase in power is possible (like resistors, capacitors and inductors, transformers are passive components), an increase in secondary voltage can only be achieved by a corresponding reduction in secondary current, and vice versa. In fact, the secondary power will be very slightly less than the primary power due to losses in the transformer. The relationships between primary and secondary voltage, current and number of turns in a transformer is as follows:

$$\frac{V_P}{V_S} = \frac{N_P}{N_S} \quad \text{and} \quad \frac{I_S}{I_P} = \frac{N_P}{N_S} \quad \text{where } N \text{ is the number of turns}$$

Another invisible friend

Like electric current, magnetic flux can't be seen — we only know that it's there by what it does. Look around your home. How many things can you see that would not be usable, if there were no electricity?

Ferromagnet – a material (such as iron, steel, or ferrite) that supports a magnetic flux

Inductor – a component that stores energy in the form of a magnetic field

Transformer – couples a.c. power from one circuit to another; can step-up or step-down voltage

Activity

1. If 6 V appears across the terminals of an inductor when the current through it changes at a constant rate from zero to 3 A in 2 seconds, what will the value of inductance be?

2. A transformer is fed from a 110 V a.c. supply. If the transformer has 660 primary turns, how many turns will be needed on the secondary to produce an output of 12 V?

Example

A transformer connected to the 220 V a.c. mains supply has 1100 primary turns and 55 secondary turns. The secondary voltage will be given by:

$$V_S = V_P \frac{N_S}{N_P} = 220 \times \frac{55}{1100} = 11\ V$$

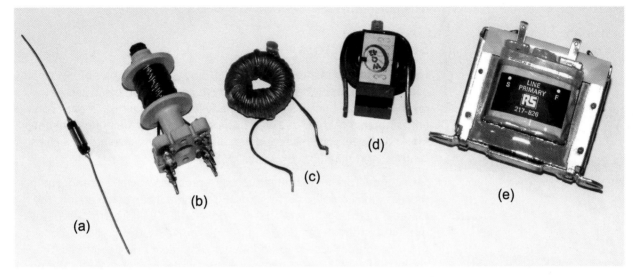

Some common inductors: (a) Ferrite cored inductor (b) Variable ferrite cored inductor (c) Toroidal ferrite cored inductor (d) Iron cored inductor (e) Transformer

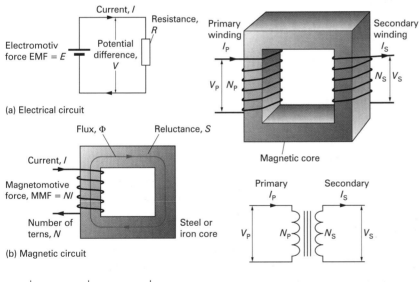

(a) Electrical circuit

(b) Magnetic circuit

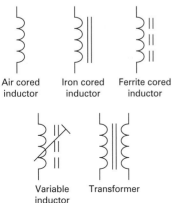

Air cored inductor
Iron cored inductor
Ferrite cored inductor

Variable inductor
Transformer

Activity

Investigate the construction of a small power transformer. Identify the main parts of the transformer and sketch a diagram showing how they are assembled.

Obtain manufacturers' data for a small step-down transformer. List and explain the specification of the component and state typical applications for it.

Just checking

* What is an inductor and what is inductance?
* What is the relationship between induced voltage, current and time?
* What is a transformer and how does it operate?
* What is the relationship between the primary and secondary voltage, current and number of turns?
* What symbols are used for showing inductors and transformers on circuit diagrams?

Diodes

Semiconductors are used in a wide variety of electrical and electronic circuits. One of the most basic forms of semiconductor has just two connections and is known as a diode. As you will see, this humble device has a wide range of applications and is available in several different forms.

Semiconductors

Like transistors and integrated circuits, diodes are examples of electronic components that we refer to generally as semiconductors. They are given this name because they are made from materials (such as silicon and germanium) that have electrical properties that are quite different from conductors (such as copper or aluminium) and insulators (such as ceramics and glass).

Semiconductors are manufactured by a process known as doping. In this process, small amounts of an impurity element are introduced into the material's lattice of atoms. The impurity element will determine whether the semiconductor material is either **N-type** or **P-type**. All diodes consist of a junction between these two types of semiconductor material. The connection to the P-type material is the **anode** while that to the N-type material is the **cathode**.

Diode operation

Diodes allow electric current to flow in one direction only. Diodes have two connections: the anode and the cathode. To make a diode conduct, the anode connection needs to be more positive than the cathode connection and then the diode is said to be conducting or forward biased. When the anode connection is more negative than the cathode connection, the diode is said to be non-conducting or reverse biased.

You can thus think of a diode as acting a bit like a switch. When forward biased, the diode acts like a closed switch (in other words it is 'on'). When reverse biased the diode acts like an open switch (in other words it is 'off'). You can see this in the diagram.

A small threshold voltage needs to be applied to a diode before the diode will actually start to conduct. The **forward voltage** for a germanium diode is approximately 0.2 V, while that for a silicon diode is about 0.6 V.

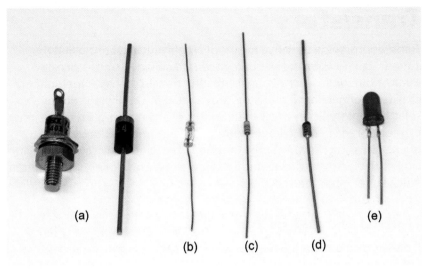

Some common diodes: (a) Silicon power rectifiers; (b) Germanium signal diode; (c) Silicon switching diode; (d) Zener diode; (e) Light-emitting diode (LED)

Diode – a semiconductor device that allows current to flow in one direction only (from anode to cathode)

N-type – semiconductor material doped with an impurity to produce an excess of electrons in the lattice of a pure semiconductor material (such as silicon)

P-type – semiconductor material doped with an impurity to produce an excess of holes (i.e. gaps into which electrons can fit) in the lattice of a pure semiconductor material (such as silicon)

Anode – the more positive terminal of a diode when conducting

Cathode – the more negative terminal of a diode when conducting

Forward voltage – the threshold voltage that needs to be applied to a diode for it to conduct (about 0.2 V for a germanium diode and 0.6 V for a silicon diode)

Diode types

Some common types of diode are shown in the photograph above. These are (left to right):

✳ power/rectifier diodes – used in power supplies to convert a.c. to d.c.

✳ signal diodes – used in radio and TV receivers and mobile phones

✳ switching diodes – used to control high voltages and currents

✳ Zener diodes – used in power supplies as an accurate voltage reference

✳ light-emitting diodes (LED) – used as indicators in a wide range of circuits.

Note that the cathodes of most signal diodes, small rectifier diodes and Zener diodes are usually marked with a stripe.

Diode (rectifier, signal diode or swiching diode)

Light-emitting diode (LED)

Light-sensitive diode (photodiode)

Zener diode

Bridge rectifier

Some diode symbols

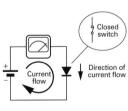

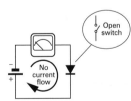

Operation of a diode

Just checking

✳ What materials are used to manufacture semiconductor diodes?

✳ How many connections does a diode have and what are they called?

✳ What is the approximate forward threshold voltage for (a) a germanium diode and (b) a silicon diode?

Transistors

Transistors provide us with a means of amplifying and controlling the current that flows in an electronic circuit. Like diodes, transistors are semiconductor devices but, unlike diodes, transistors have three, rather than two connections. Transistors are capable of amplifying current – a small current (typically a few thousandths of an amp) applied to the input can control a very much larger current (several amps in the case of a power transistor) flowing at the output. These handy devices are used in almost all electronic circuits, either as individual components or as part of much larger integrated circuits.

Transistor construction

Conventional **bipolar junction transistors (BJT)** comprise either NPN or PNP junctions of semiconductor material (usually silicon). The junctions are formed on a single slice of silicon by diffusing impurity elements through a photographically reduced mask. We call them bipolar because they use both positive and negative charge carriers.

The three connections on a BJT are referred to as the base, emitter and collector. Inside a BJT there are two semiconductor junctions; the base-emitter junction and the base-collector junction.

Transistor operation

In normal operation, the base-emitter junction of a BJT is forward biased and the collector-base junction is reverse biased. The base region is, however, made very narrow so that carriers are swept across it from emitter to collector. Thus, only a relatively small current flows in the base. To put this into context, the current flowing in the emitter circuit is typically 100 times greater than that flowing in the base. The direction of conventional current flow is from emitter to the collector in the case of a PNP transistor, and collector to the emitter in the case of an NPN device.

The equation that relates current flow in the collector, base, and emitter currents is:

$$I_E = I_B + I_C$$

where I_E is emitter current, I_B is base current, and I_C is collector current (all expressed in the same units).

So, if a transistor operates with a base current of 1 mA (0.001 A) and a collector current of 99 mA (0.099 A), the transistor must have an emitter current of $(1 + 99) = 100$ mA.

In the simple amplifier circuit shown (where the emitter is common to the input and output circuits), the small input current is applied to the base and the much larger output current flows in the collector circuit. This circuit is known as a **common emitter amplifier.** If the base current doubles, the collector current will also double. If the base current falls by 50%, the collector current will also fall by 50%. If the base current falls to zero, the collector current will also fall to zero. The transistor is said to have current gain, β, which is calculated from:

$$\beta = \frac{I_C}{I_B}$$

In control

Diodes are useful because they allow current to flow in one direction and stop it flowing in the other. Another useful device would be one that could actually control the current – a small current flowing into the device controlling a much larger current flowing out of it. Can you think of where this might be useful?

Transistor – a semiconductor device with three terminals that can be used as a current amplifier

Bipolar junction transistor (BJT) – a transistor with two junctions (either NPN or PNP) that operates with both positive and negative charge carriers

Common emitter amplifier – BJT used as an amplifier with emitter connection common to both input and output

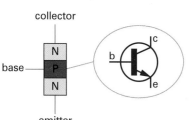

(a)NPN bipolar junction transistor (BJT)

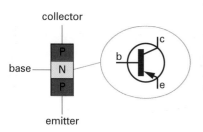

(b)PNP bipolar junction transistor (BJT)

So, if a base current of 10 mA (0.01 A) produces a collector current of 1.5 A, the transistor must have a common emitter current gain (β) of 150. Typical values range from about 30 to around 250.

Transistor types

Some common types of transistor are shown in the photograph below. These are (left to right):

* small signal transistors – used in amplifiers, radio and TV receivers

* switching transistor – used in switching applications where current and voltage has to be controlled

* power transistors – used in power supplies, audio amplifiers, automotive applications, etc.

* phototransistor – used for light sensing and optical data transmission

* radio frequency transistor – used in very high frequency applications.

Note that many different package styles are used for transistors. Connections can be obtained from manufacturers' data sheets.

Activity

Construct a simple single-stage transistor amplifier. Measure the input and output voltage and calculate the voltage gain provided by the amplifier stage. Can you suggest an application for the amplifier circuit?

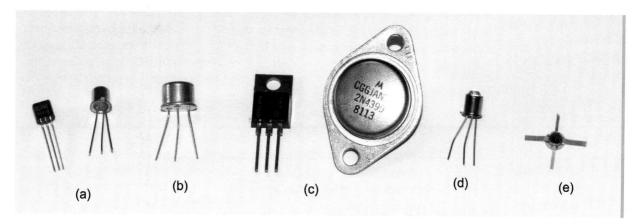

(a)   (b)   (c)   (d)   (e)

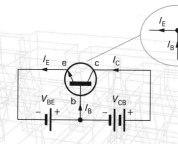

(a) Current flow in an NPN bipolar junction transistor

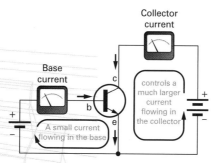

Some common transistors: a) Small signal transistors b) Switching transistor c) Power transistors d) Phototransistor e) Radio frequency transistor

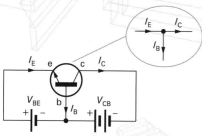

(b) Current flow in a PNP bipolar junction transistor

Operation of a transistor

Just checking

* How many connections does a bipolar junction transistor (BJT) have and what are they called?

* What symbols are used for NPN and PNP transistors and how do they differ?

* What is the relationship between the currents that flow in a BJT?

* How is the current gain for a common emitter amplifier calculated and what is a typical value?

* What different types of transistor are available and what are they used for?

Integrated circuits

The first integrated circuit (or 'chip') was developed in the 1950s by Jack Kilby of Texas Instruments and Robert Noyce of Fairchild Semiconductor. Today, chips are used in nearly all electronic devices, including mobile phones, MP3 players, audio, radio and video equipment, computers, domestic equipment (such as washing machines and microwave ovens), cars and aircraft.

Integrated circuit construction

Considerable savings of cost and space can be made by manufacturing all of the components required for a particular circuit function (such as an amplifier or digital logic circuit) on a single slice of silicon. The resulting integrated circuit may contain as few as 10 or several million resistors, capacitors, diodes and transistors.

Integrated circuits are manufactured using a series of stencil-like patterns (called masks). These are used to protect parts of a chip from ultraviolet light during a fabrication process called photolithography. The various doped regions within the chip are built up in a series of layers. The final chip is cut from the die on which it is produced and then mounted inside a ceramic or plastic package. It takes an average of 200 people, working full time for two years to design, test, and prepare a new chip design for fabrication!

Scale of integration

The scale of integration of an integrated circuit chip is often quoted in terms of the number of electronic components that it contains. A simple integrated circuit (such as an operational amplifier) may have less than 100 components, whereas the microprocessor used in a computer may have several million.

Technology	Abbreviation	Number of active devices/logic gates
Small-scale integration	SSI	Up to 100 electronic components per chip
Medium-scale integration	MSI	From 100 to 3,000 electronic components per chip
Large-scale integration	LSI	From 3,000 to 100,000 electronic components per chip
Very large-scale integration	VLSI	From 100,000 to 1,000,000 electronic components per chip
Ultra large-scale integration	ULSI	More than 1 million electronic components per chip

Integrated circuit types

Integrated circuits can be divided into two general classes: **linear** (analogue) and **digital**, according to the particular kind of signals and applications that the chip is designed to work with. Typical examples of linear integrated circuits are operational amplifiers, whereas typical examples of digital integrated are logic gates (see opposite). A number of devices bridge the gap between the analogue and digital world.

Getting it together

Integrated circuits make it possible to pack highly complex electronic circuits into a very small space. This saves cost and size, and many items of electronic equipment that we all take for granted would not be possible without them. Try to list all the items that you have at home that you think use integrated circuits.

Personal learning and thinking skills

Make an A3-sized poster that lists and briefly describes the main stages used in the manufacture of an integrated circuit.

Integrated circuit
– semiconductor device with a large number of components on a single chip of silicon

Linear integrated circuit
– integrated circuit designed for analogue (linear) applications

Digital integrated circuit
– integrated circuit designed for digital applications

Hybrid integrated circuit
– integrated circuit that uses both analogue and digital technology

Such devices (often referred to as **hybrids**) include analogue to digital converters (ADC), digital to analogue converters (DAC), and timers.

Some common types of integrated circuit are shown in the photograph. These are (left to right):

Some common integrated circuits:
a) Operational amplifier b) Timer c) Voltage regulator d) Logic gate e) Memory (ROM)
f) Consumer integrated circuit

✳ **operational amplifiers** – designed primarily for linear operation, forming the fundamental building blocks of a wide variety of linear circuits such as amplifiers, filters and oscillators

✳ **timers** – designed primarily for generating signals that have an accurately defined time interval, such as providing a delay or determining the time between pulses

✳ **voltage regulators** – used in power supplies to maintain a constant output voltage, when the input voltage or the load current changes over a wide range

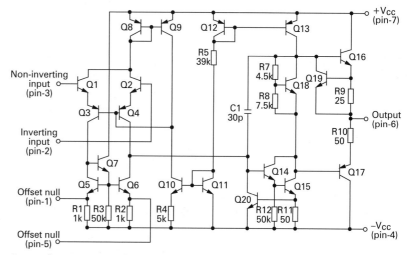

Circuit of a 741 integrated circuit

Activity

Use the internet (or other information sources) to find out how integrated circuits are manufactured. Hint: You can download an excellent booklet called 'From sand to circuits' from Intel's website (www.intel.com).

Obtain manufacturers' data for an LM386N integrated circuit and use it to answer these questions:

a) What is the intended application for the chip?

b) Over what range of supply voltages will the chip operate?

c) What voltage gain does the chip provide?

Also sketch a labelled diagram showing the pin connections for the chip (viewed from above).

✳ **logic gates** –performing logical functions such as AND, OR, NAND and NOR (more of this later)

✳ **memories** – used to store digital information and include the random access memories (RAM) and read only memories (ROM) used in computers

✳ **consumer integrated circuits** – such as audio amplifiers, radio receivers, etc.

Just checking

✳ What are the main advantages of integrated circuits?

✳ How are integrated circuits manufactured?

✳ What are the main types of integrated circuit?

✳ What do the terms SSI, MSI, LSI, VLSI and ULSI mean?

Analogue signal – a signal that varies continuously over a wide range of values

Digital signal – a signal that exists in (usually two) discrete steps or voltage levels

Operational amplifier – a type of integrated circuit designed for analogue (linear) applications

Voltage

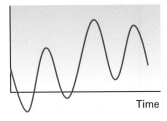

Time

(a) An analogue signal

Voltage

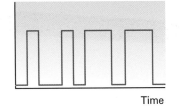

Time

(b) A digital signal

Analogue circuits

Analogue signals involve smoothly changing voltage levels, rather than the continuous rapid switching between just two voltage levels in digital circuits. Analogue circuits are found in a wide range of electronic applications including radio and TV receivers and audio amplifiers.

Analogue signals

Because analogue circuits can represent a continuous range of values, they are used in applications where signal levels need to vary over a wide range. For example, the sound produced by an orchestra can vary in intensity from barely audible to the loudest crescendo. An analogue circuit would be needed to pick up the sound from a microphone and amplify it to a level at which it could be recorded.

The voltages and currents in an analogue circuit vary continuously just like the signals in the real world that they represent. **Digital signals**, by contrast, involve switching on and off rapidly between two voltage levels. They use digital codes to represent the voltages and currents that would appear in an analogue circuit.

Operational amplifiers

Operational amplifiers are analogue integrated circuits designed for linear amplification that offer near-ideal characteristics (virtually infinite voltage gain and input resistance coupled with low output resistance and wide bandwidth).

They can be thought of as universal 'gain blocks' to which external components are added to define their function within a circuit. By adding two resistors, you can produce an amplifier having a precisely defined gain. Alternatively, with three resistors and two capacitors you produce a filter that will amplify only low or only high frequencies. From this you might begin to suspect that operational amplifiers are really easy to use. The good news is that they are!

Look at the symbol for an operational amplifier. The component (an integrated circuit) has two inputs and one output, and two supply connections (positive and negative). In fact, you sometimes don't show the supply connections because it's often clearer to leave them out of the circuit and just assume they are there!

In the diagram, one of the inputs is marked '−' and the other is marked '+'. These polarity markings have nothing to do with the supply connections – they indicate the overall phase shift between each input and the output. The '+' sign indicates zero phase shift (in which case the output and input signals rise and fall together) while the '−' sign indicates a phase shift of 180° (in which case the output signal will be turned upside down). The '−' input is often referred to as the inverting input. Similarly, the '+' input is known as the non-inverting input.

Most, but not all, operational amplifiers require a symmetrical supply of typically between ±5 V and ±15 V. This arrangement allows the output voltage to swing both positive (above 0 V) and negative (below 0 V). Note that you usually have two separate supplies: a positive supply and an equal, but opposite, negative supply. The common connection to these two supplies (i.e. the 0 V rail) acts as the common rail.

The characteristics of most modern integrated circuit operational amplifiers come very close to those of a perfect amplifier, as you can see from the table below.

Parameter	Ideal	Real
Voltage gain	Infinite	100,000
Input resistance	Infinite	100 MΩ
Output resistance	Zero	20 Ω
Bandwidth	Infinite	2 MHz

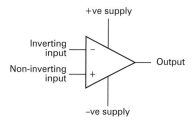

Activity

Obtain manufacturers' data for a 741N integrated circuit and use it to answer the following questions:

a) What is the intended application for the chip?

b) Over what range of supply voltages will the chip operate?

c) What voltage gain does the chip provide?

Also sketch a labelled diagram showing the pin connections for the chip (viewed from above).

Example analogue circuit - an audio amplifier

The audio amplifier uses an LM386 consumer integrated circuit. The LM386 was designed specifically for use in battery-operated audio equipment and radio receivers and provides a voltage gain of around 200. This makes it ideal for operation as a low-power audio amplifier, where its output can be fed to a small loudspeaker.

Capacitor C1 removes any d.c. voltage that might be present at the input and VR1 (a variable resistor) acts as a volume control. The signal input to IC1 appears between pins 3 and 2 and the output appears at pin 5. The output is coupled to the loudspeaker through C4 which prevents d.c. voltage from reaching the loudspeaker while, at the same time, coupling the a.c. signal to it. A double-pole single-throw (DPST) switch is used as an on/off switch connected in series with the 9 V battery.

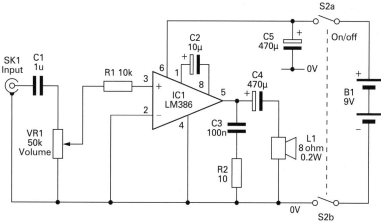

Just checking

* What are the characteristics of analogue signals?
* In what typical applications are analogue signals used?
* What is an operational amplifier and what does it do?
* What is an audio amplifier and over what frequency range does it operate?

Logic gate – a digital circuit that performs a logic function, such as AND, OR, NAND and NOR

Bistable circuit – a digital circuit that retains its logic state indefinitely or until set or reset. A bistable circuit acts like a simple memory, storing a logic 0 or logic 1 condition

CMOS – a logic family based on complementary metal oxide semiconductor technology

TTL – a logic family based on transistor-transistor logic

Digital circuits

Digital circuits are based on straightforward on/off switching, but that doesn't mean that they are only suitable for use in simple applications. You will find them in computers and a host of other applications that are programmed by software. They are also used in many other circuits where logical decisions need to be made. You need to understand how digital circuits operate and know how they are used in simple electronic circuits.

Digital signals

The signal voltages in an analogue circuit can take any value within a given range and small changes can often be important. By contrast, in nearly all digital circuits the signal voltages can only take one of two possible values. These discrete steps are usually referred to as logic 0 and logic 1 and they are often represented by fixed voltages of 0 V and +5 V respectively. Provided they do not stray outside the range of voltages that we assign to logic 0 and logic 1, any small changes are completely ignored in a digital circuit.

Logic gates

The British Standard (BS) and American Standard (MIL/ANSI) symbols for some basic logic gates with two inputs are shown in the diagram, together with their truth tables. The action of each of the basic logic gates is summarised below. Although we've shown only two inputs, the basic gates (AND, OR, NAND and NOR) are commonly available with up to eight inputs.

* **AND gates**. AND gates will only produce a logic 1 output when all inputs are simultaneously at logic 1. Any other input combination results in a logic 0 output.

* **OR gates**. OR gates will produce a logic 1 output whenever any one, or more, inputs are at logic 1. Putting it another way, an OR gate will only produce a logic 0 output when all of its inputs are simultaneously at logic 0.

* **NAND gates**. NAND gates will only produce a logic 0 output when all inputs are simultaneously at logic 1. Any other input combination will produce a logic 1 output. A NAND gate, therefore, is nothing more than an AND gate with its output inverted! The circle shown at the output denotes this inversion.

* **NOR gates**. NOR gates will only produce a logic 1 output when all inputs are simultaneously at logic 0. Any other input combination will produce a logic 0 output. A NOR gate, therefore, is simply an OR gate with its output inverted. A circle is again used to indicate inversion.

* **Bistables**. The output of a bistable has two stable states. It can be either set (in which case its output will be logic 1) or it can be reset (in which case its output will be logic 0). Once set, the output of a bistable will be preserved until it is reset. A bistable thus acts like a simple form of memory, as it will remain in its latched state (set or reset) until commanded to change its state (or until the supply is disconnected). Various forms of bistable are available, including R-S, D-type and J-K types.

Truth tables are frequently used to describe the operation of digital logic circuits. The truth table shows all of the possible input states (there will be four possible input states for a two-input logic gate, eight for a three-input logic gate, sixteen for a four-input logic gate, and so on) together with the resulting output state. The inputs are usually labelled A, B, C, and so on, while the output is usually shown as X or Y.

Logic families

Digital integrated circuit devices are often classified according to the semiconductor technology used in their manufacture. The logic family to which a device belongs determines its operational characteristics, such as the voltage levels used to represent logic 0 and logic 1, power consumption, speed and immunity to noise.

The two basic logic families are **CMOS** (Complementary Metal Oxide Semiconductor) and **TTL** (Transistor Transistor Logic). Each of these families is then further sub-divided. The most common family of TTL logic devices is known as the 74-series. Devices from this family are coded with the prefix number 74. Sub-families are identified by letters that follow the initial 74 prefix, such as 74ALS, 74HCT, and so on.

CMOS devices use significantly less power than TTL devices. They also operate over a much wider range of supply voltages (typically from +3 V to + 18 V). Because of this, CMOS devices are often used for portable, battery operated equipment. TTL devices, on the other hand, require a stable (well regulated) +5 V supply but they are capable of operating at much faster speeds than equivalent CMOS devices.

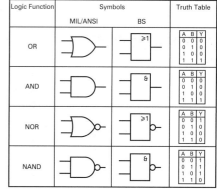

Symbols and truth tables for some basic logic gates with two inputs (A and B are the two inputs and Y is the output)

Activity

The simple theft alarm can be silenced by simply replacing the link that has been broken. A much better alarm circuit would use a bistable element to 'remember' that the alarm had been triggered. Investigate the use of bistable circuits. Modify the intruder alarm incorporating a bistable, so that the alarm will continue to sound even after a broken link has been reinstated.

Personal learning and thinking skills

Use library and internet resources to investigate the operation and application of simple R-S and D-type bistable circuits. Present your findings in the form of a brief PowerPoint presentation to the rest of the class. When making your presentation, allow some time for questions.

Example digital circuit – a theft alarm

A simple example of a digital circuit is shown in the diagram: an alarm that uses wire links to protect an item of property (such as a bike) against theft. The four shorting links (LKA to LKD) are used to hold the inputs of the four-input OR gate (IC1) down to logic 0. If any link is broken, the respective input will go high (to logic 1) due to the pull-up resistors (R1 to R4). If any one (or more) of the inputs goes to logic 1, the output of IC1 (at pin-1) will also go to logic 1. This will cause current to flow into the base of TR1 (through R5). TR1 will then conduct heavily, acting as a switch, which causes current to flow through the audible transducer (PZ1) and also through the LED (D1) and R6 connected in parallel with it. The alarm can be interrupted by means of the single-pole single-throw switch (SPST) switch, S1.

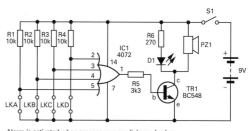

Alarm is activated when any one or more links are broken

A simple theft alarm

Just checking

* What are the characteristics of digital signals?
* What does a logic gate do and how are logic levels represented?
* What does a bistable do?
* What basic logic gates are available and what are their symbols?
* What is the difference between CMOS and TTL technology?

Test equipment and measurements

Some basic test equipment is required to test and carry out measurements on electronic circuits. This includes **multi-range meters** (capable of measuring voltage, current and resistance), logic probes, logic pulsers, and **oscilloscopes**. You will have an opportunity to use some of these items when you build, test and carry out fault-finding on some simple electronic circuits.

Multi-range meters

Multi-range meters provide either analogue or **digital** indications of voltage, current and resistance. They are usually battery-powered and, thus, readily portable. Controls and adjustments are extremely straightforward and a connection to the circuit under test is made via a pair of test leads fitted with probes or clips. Most multi-range meters will allow you to measure d.c. and a.c. voltage, d.c. and a.c. current, and resistance. Some meters also have additional ranges for checking batteries, diodes, transistors and capacitors. Typical applications for multi-range meters include checking supply voltages, as well as the voltages at test points (such as the connections to transistors, diodes and integrated circuits).

Logic probes

Logic probes are ideal for checking digital circuits. They are much more convenient than a digital multi-range meter when checking logic circuits because they indicate logic 0 and logic 1 states directly. Logic probes consist of a small hand-held probe fitted with LEDs to indicate the logical state at the probe tip. Most logic probes incorporate a pulse stretching circuit that elongates pulses of very short duration, so that a visible indication is produced on a separate pulse LED. Logic probes invariably derive their power supply from the circuit under test and are connected by means of a short length of twin flex fitted with insulated crocodile clips.

Logic pulsers

You may sometimes need to generate the logic levels required by a digital circuit. A permanent logic level can easily be generated by pulling a line up to + 5 V via a pull-up resistor, or by temporarily tying a line to 0 V. On other occasions, you may need to generate a momentary pulse rather than a permanent logic state. You can do this with the aid of a logic pulser. This momentarily forces a logic level change into a circuit and avoids the need to disconnect or de-solder any of the components. The polarity of the pulse (produced at the touch of a button) is adjusted so that the node under investigation is momentarily forced into the opposite logical state. During the period before the button is depressed, and for the period after the pulse has been completed, the probe tip adopts a tri-state (high resistance) condition and so does not permanently affect the logical state. Like logic probes, pulsers derive their power supply from the circuit under test.

A typical digital multi-range meter with a.c., d.c. and resistance ranges

Oscilloscopes

An oscilloscope allows you to display time-related voltages (or waveforms). Oscilloscopes are often used for displaying analogue signals and detecting noise and distortion that may be present, as well as investigating voltages and general waveform shapes. An oscilloscope makes it easy to measure the periodic time, frequency and peak-peak voltage of an AC signal.

Oscilloscopes incorporate a timebase circuit that determines the horizontal (time) scale. You can adjust the display on the screen to show a greater or lesser number of cycles of the wave. The relationship between frequency and periodic time is given by:

$$f = \frac{1}{t}$$

where f is the frequency in Hertz (Hz) and t is the periodic time in seconds (s).

Example

A waveform has a frequency of 50 Hz. Its period will be given by:

$$t = \frac{1}{f} = \frac{1}{50} = 0.02 \text{ s (or 20 ms)}$$

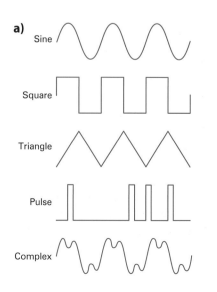

A typical dual-beam bench oscilloscope

A typical logic probe incorporating a pulse stretching circuit

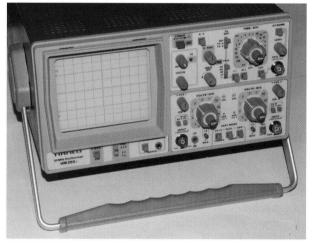

a)
Sine
Square
Triangle
Pulse
Complex

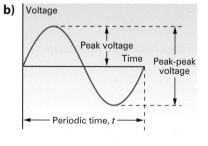

b)
Voltage
Peak voltage
Time
Peak-peak voltage
Periodic time, t

Waveforms:
a) some common waveforms
b) peak and peak-peak voltage and periodic time for a sine wave

Electronic circuit construction

Various methods are used for building electronic circuits. The method chosen for a particular application depends on a number of factors, including the available resources and the scale of production. Techniques used for large-scale electronic manufacture generally involve fully automated assembly, using equipment that can produce complex circuits quickly and accurately, at low cost with minimal human intervention. On the other extreme, if only one circuit is to be built, a hand-built prototype is much more appropriate.

Construction methods

Various methods are used to construct electronic circuits, including:

* point-to-point soldered wiring
* breadboard construction
* printed circuit construction
* surface mounting.

Point-to-point wiring

Point-to-point wiring involves linking electrical connections using short lengths of insulated wire soldered to the pins and tags (as appropriate). Tag strips and terminal blocks are also used. This method of construction is now considered obsolete with the advent of miniature components, printed circuit boards and integrated circuits, but is still used for some simple applications (for example, the wiring inside a portable electric drill).

Breadboard construction

This 'solderless' construction technique is often used for assembling and testing simple prototype circuits. The advantage is that changes can be quickly and easily made to a circuit, and all of the components can be re-used. Disadvantages are that it is unsuitable for permanent use and for complex circuits.

An example of breadboard construction

Matrix board construction

This low-cost technique avoids the need for a printed circuit, but is generally only suitable for one-off prototypes. The matrix board consists of an insulated board into which a matrix of holes is drilled, with copper tracks arranged as strips on the reverse side of the board. Component leads are inserted through the holes and soldered into place. Strips (or tracks) are linked together with

Matrix board construction

Matrix board construction – a method of constructing an electronic circuit using a circuit board that has strips of copper tracks with holes arranged in the form of a matrix

Printed circuit construction – a method of constructing an electronic circuit in which the electrical connections and contacts are made using copper tracks and pads that are 'printed' on the surface of an insulating board.

Surface mounting – a technique used in the automated assembly of electronic circuits. Surface-mounted components do not have conventional leads and are soldered into place using a solder paste, applied to their mounting pads on the surface of the printed circuit board

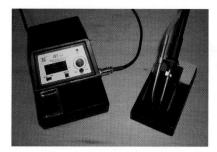

A temperature controlled soldering station

A printed circuit board

short lengths of tinned copper wire (inserted through holes in the board and soldered into place on the underside of the board). Tracks can be broken at various points as appropriate. Note that a suitable rating for a soldering iron for light electronic work (matrix board and small printed circuit boards) is typically between 15 W and 25 W. Larger soldering irons (particularly those that are not temperature-controlled) may cause damage to tracks, pads and components.

The advantage of matrix boards is that they avoid the need for a printed circuit board (which may be expensive and take some time to design). Disadvantages are that it is usually only suitable for one-off production and the end result is invariably less compact than a printed circuit board.

Printed circuit board construction

This is ideal for volume manufacture of electronic circuits where speed and repeatability of production are important. Depending on the complexity of a circuit, various types of printed circuit board are possible.

The most basic form of printed circuit has copper tracks on one side and components mounted on the other. More complex printed circuit boards have tracks on both sides (referred to as 'double-sided') while boards with up to four layers are used for some of the most sophisticated and densely packed electronic equipment (for example, computer motherboards).

Modern printed circuits often use leadless **surface-mounted** components. These circuits are designed for automated soldering directly to pads on the surface of a printed circuit board. This makes it possible to pack in the largest number of components into the smallest space but, since the components require specialised handling and soldering equipment, it is not suitable for one-off construction for hand-built prototypes.

Activity

Construct simple electronic circuits using each of the following techniques:

a breadboard construction

b matrix board construction

c printed circuit board construction.

In each case, assemble and test your circuit and check that it operates correctly. Which of the three techniques do you prefer and why do you prefer it?

Just checking

* What different techniques can be used to construct an electronic circuit?
* What are the advantages/disadvantages of each method of electronic circuit construction?

Fault finding

Electronic circuits can fail for a number of reasons. It can be due to poor design or improper manufacture, but sometimes electronic components fail for no apparent reason. Since many electronic circuits are complex and involve hundreds of individual components, fault finding might appear to be challenging. In fact, provided you use a logical approach, know how circuits work and can make effective use of test equipment, you should be able to quickly find the cause and put it right.

Fault-finding procedure

The first stage is to ensure that the equipment really is defective! This may sound rather obvious but, in some cases, a fault may simply be due to poor adjustment or misconnection. Where several items of equipment are connected together, it may be necessary to first pinpoint which is faulty. For example, consider a computer system, which the user simply states 'doesn't work'. The fault could be almost anywhere: computer, display, keyboard, mouse, printer or any one of several connecting cables.

The second stage is to gather all relevant information. This process involves asking questions like these.

* In what circumstances did the equipment fail?
* Has the equipment operated correctly before?
* Exactly what has changed?
* Is the fault intermittent or is it present all of the time?
* Has there been a progressive deterioration in performance?
* Has the system operated in similar circumstances without failure?
* If the fault is intermittent, in what circumstances does it happen?
* Is it possible to predict when the fault will occur?
* What parts of the equipment are known to be working correctly?
* Is it possible to isolate the fault to a particular area?
* Is the fault a known 'stock fault' – has the fault been documented elsewhere?

These questions are crucial, particularly when you have no previous experience of the equipment. Coupled with knowledge of what you expect it to do, the answers will help decide how and when the fault happened and what went wrong.

Once you have analysed the information, the next stage involves making measurements to identify the faulty component or components. A fuse may have blown or a component may have failed. Once the failed component has been identified, it can be removed and replaced and the equipment should work again.

Electrical safety precautions are essential when working on live equipment. Use insulated test leads and prods and avoid contact with the circuit under investigation. A good technique is to use only one hand when taking measurements – keeping the other safely out of the way. This can help avoid an electric shock. Most battery-operated equipment is safe to work on but great care should always be taken with mains operated equipment, as the voltages and currents can be lethal. If in doubt, consult your teacher before doing anything!

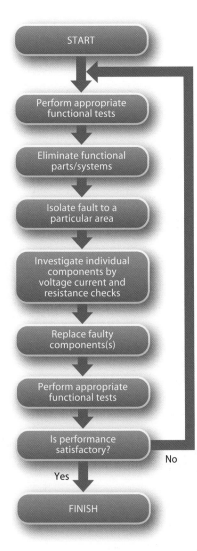

Fault finding procedure

Fault-finding charts and voltage tables

It is always useful to obtain a circuit diagram and service information for the equipment before starting work. Also, fault finding can be aided by **fault-finding charts** and **voltage tables** supplied by manufacturers. The former provide a step-by-step procedure for checking and eliminating functional parts, while the latter provide normal working voltages (and sometimes also currents) measured at a number of **test points.**

Fault-finding chart – a diagram showing logical steps to find a fault on an electronic circuit quickly and easily

Voltage table – normal working voltages for a circuit, often included in service information from manufacturers.

Test point – a point in a circuit where test voltages and/or currents are taken

Activity

Carry out a range of measurements on at least three simple electronic circuits and use these, together with test voltages and fault location charts (as appropriate) to locate simple faults

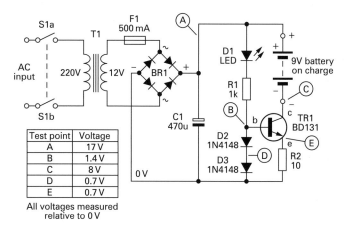

Test point	Voltage
A	17 V
B	1.4 V
C	8 V
D	0.7 V
E	0.7 V

All voltages measured relative to 0 V

Circuit diagram of a charger for recharchable batteries with typical test-point voltage table

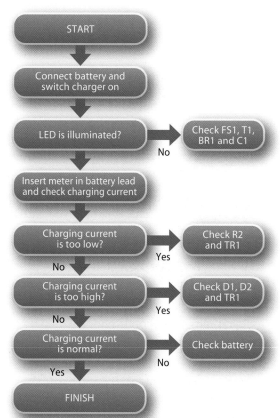

Simplified fault finding chart for a battery charger

Just checking

* What is a fault-finding chart and how does it help you to locate a fault?
* What is a voltage table and how does it help you to locate a fault?
* What is a test point and how can it be used in fault finding?
* What precautions must you observe when making measurements on live circuits?

Unit 5 Assessment Guide

In this unit, you will investigate how electrical and electronic circuits are constructed and tested in laboratories and workshops. The starting point is circuit design based on electrical and electronic principles, followed by the selection of components and construction using safe working practices.

Time Management

Manage your time well as this unit has a number of different components that will have to be researched. Ensure that you keep your work safe and that any work in electronic format has a secure and safe backup.

Be well organised. This is your chance to show that you are an independent enquirer, creative thinker, and a self manager and therefore will contribute towards achievement of your Personal learning and thinking skills.

Plan ahead for your work experience by making a checklist of things that you will need to investigate so that you make the most efficient use of your time

Useful Links

Make good use of your work experience to find out as much as possible about issues that are relevant to your coursework. You should try to meet with people who work in the design and development departments of an electronics company. Talk to them about their jobs and make a note of what they say.

Use an electronic component supplier's catalogue (hard copy, CD ROM or web based) and wall charts to help you with assessment focus 5.2.

Things you might need

Your work should be presented as an A4 process portfolio and should be in e-format. Your teacher should give you access to the required software to enable the correct presentation.

A digital camera or mobile phone would be useful, so that you can record evidence of practical activities.

Remember that you will have to carry out workshop activities in order to complete assessment foci 5.3 and 5.4. Make sure that you obtain witness statements from your supervisor and photographs showing you working correctly and safely.

Remember to maintain a focus on why the electronics industry is continually coming up with new products. It is because it employs engineers who can design, build and test circuits.

How you will be assessed		
What you must show that you know:	**Guidance:**	**To gain higher marks:**
That safe working practices must be followed when using electrical equipment and constructing circuits in a workshop or laboratory. Also, to design a circuit you need to know about the theory of electrical and electronic principles. *Assessment focus 5.1*	* Explain the safe working practices which must be used when working on electrical and electronic circuits. * Describe how electric shock can be prevented when working with high voltage circuits. * Describe how to carry out safe soldering.	* You need to provide evidence that you adopted safe working practices when constructing, testing and fault finding circuits (assessment focus 5.3) * You need to measure the current, voltage and power in a simple DC circuit and compare these with calculated values. * You must calculate the ratings for a fuse and a circuit breaker in an AC circuit.
How to recognise and select components used in electrical and electronic circuits. *Assessment focus 5.2*	* From a box of mixed parts identify six different types of component used in electrical and electronic circuits. * For each component use reference data to establish its properties e.g. the value of a resistor indicated by colour bands.	* You must read an electrical/electronic circuit and identify a number of different components in it. * Using a components supplier's catalogue and the circuit diagram you must: * select components * write out a requisition order for the components
How to construct an electronic circuit consisting of a number of basic components and to understand how it works. *Assessment focus 5.3*	* Use hand tools to construct a circuit. * Obtain witness statements and take photographs to prove that you carried out the practical work effectively. * Describe what the circuit does.	* You must describe how the circuit works and the function of each component * You should use electrical/electronic theory to support your description of how the circuit functions.
How to use test equipment in order to check if electronic circuits are working correctly and if not, to determine why they are faulty. *Assessment focus 5.4*	* Visually inspect a circuit board to find any construction faults. * Carry out voltage and current measurements on a circuit to check that it is working properly. * Use test equipment to locate faults in a circuit. * Obtain witness statements and take photographs to prove that you carried out the practical work effectively.	* You must make test measurements and compare your readings with the values given in the circuit specification. * You must be able to identify and explain faults in a circuit by carrying out a logical test process.

Introduction

Almost everything you use in today's world has been designed or made by engineers. All of the products that you use on a daily basis – from washing machines to buses, to knives and forks – are engineered, and many are manufactured in the UK. It is very exciting to be involved with a company producing items you see every day, the noise of the machines working, the sense of producing something useful and even the smell of engineering lives with you for life.

Most manufacturers have people who understand modern technology, computers and manufacturing methods. These people are engineers who to try to gain an advantage over competitors. These engineers are responsible for manufacturers producing high quality, high value-added goods and services.

As well as working in a team producing a quantity of the same product, in this chapter you will learn about production planning and scheduling, and will look at the need for setting up tools and equipment, including computer numerical control (CNC), and preparing materials ready for production. You will learn about applying quality control during production, such as statistical process control, and find out about recording and interpreting the resulting data.

How you will be assessed

This unit will be assessed by your tutor who will set an assignment for you to complete. Because of the nature of this unit you will need to work as part of a production team to manufacture a quantity of the same, engineered product. As such, you will be assessed through the theme of the work of a manufacturing engineer, including planning, scheduling, setting machines and quality control. Although working in a team, the evidence that you produce for assessment must always be your own work.

After completing this unit you should be able to achieve the following outcomes:

1. Participate effectively in a production team and describe your own role in the team, with an indication of your strengths and weaknesses.

2. Describe the essential production information found in product drawings and a specification for a given engineered product.

3. Produce a production plan and schedule for the manufacture of a quantity of the same engineered product.

4. Set up and use CNC tools and equipment to manufacture a quantity of the same simple engineered product safely, using pre-prepared materials, maintaining a clean and tidy working environment. Comment on the success of this activity.

5. Use quality control techniques correctly, including statistical methods, to establish whether or not a sample of engineered products conform to the standards specified.

Thinking Points

Working in teams is present in every day of our lives and includes not just our home life, but our working life. Obviously we can think of sports teams – football or a formula one team – if you work in engineering you are most likely to work in some form of team or other. Take a look at the job adverts in either a local or national paper and establish what departments the vacancies exist in. Think of the teams they would be working in. Alternatively, if you have visited an engineering company, try to think back and remind yourself of the teams there.

Why is it important to record what goes on in team meetings? The agenda for a meeting outlines what is to be covered in the meeting and its order: why is this important? How do you think you would deal with conflict in a meeting or on the production line?

Do you think it is fair that an operator on a machine should guess how to make a component? Is it better that they are supplied with a range of information to help this? Think of all the things that they need to know about when they are making a component: where does all of this information come from? What do you think should go into a production plan and a production schedule?

Obviously, once manufacture has been planned, machining is carried out. Think about some of the machines that can be used for this: especially those with computer control. Have you ever thought about why some materials can be machined easier than others? This is an important aspect when machining components, as is the way we check to see if what we produce is meeting the required standard. Think about what it is like if parts don't fit together properly. A quality product will always fit together well.

An introduction to manufacturing techniques

All the products that you use on a daily basis, from computers to coffee pots, are engineered. Many are manufactured in the UK in a vibrant sector that contributes a great deal to the national economy. Most UK manufacturers have turned to modern technology and manufacturing methods to try to gain an advantage over their global competitors – and more often than not now produce high quality, high-value added goods and services.

what manufacturing techniques do you know about?

Visit the engineering workshop at your school or college and make a list of the manufacuring techniques you will find there. It is likely that you will base this on the machinery you find. Ask the people in the workshop about the techniques used when making products. When you get back to the classroom, think about these techniques: what do you really know about them?

So what will you do in this unit?

In this unit you are going to get involved in engineering manufacturing activities, by producing, as part of a **team**, a quantity of the same product. You will learn about production planning and scheduling and be responsible for setting up tools and equipment, including computer numerical control (CNC), and preparing materials ready for production. In line with modern industrial practice, you will learn how to apply quality control during production and be able to record and interpret the resulting data.

Working in teams

A team is a group of two or more people working together towards a common goal. It could be said that, in the beginning, people come together as a group of individuals and only when each other's strengths are used is the group operating as a team.

A team meeting

When you think of teamwork, you may immediately think of a sporting team – you may even support a football team. In manufacturing, teams exist as, for example, production teams, service teams or management teams. The usual and most appropriate place to find a team is one set up for shop-floor production.

How does a team develop?

A team's development may be viewed as having four different stages. It is important to realise this when setting up teams to work in manufacturing organisations.

Personal learning and thinking skills

Apart from a sporting team, like a football team, think of a situation where a team is in operation. Make a list of what you think each member does. If you were working in a team, what sort of team player would you be? Would you be an 'action' type member, a 'people' type member or a 'thinker/organiser' type member? You will need to think about what each of these do. Make some notes about each to help you identify which type you might be.

Team – people working together, but more than just a group

Roles – specified tasks or functions which would normally be divided up within the team

Responsibility – being answerable, in the first place to the other team members

These are the four stages.

* The **forming** phase, when the team is still only a collection or group of individuals dealing with basic procedures and the atmosphere is often artificially polite. Members show enthusiasm in the new team, but know-how and ability is low.

* The **storming phase**, when the team members begin to experiment and flex their muscles. Relationships become stormier, both between members and between the team and other groups, and members often do not like procedures. As the team struggles to find the best way to work together, members may have a temporary lapse in commitment.

* The **norming phase**, where the team is beginning to achieve well. It has developed its own way of working that is producing results.

* The **performing phase**, where the team is working well together. It deals with change in an open and flexible way, constantly challenging itself but avoiding damaging conflicts. Development of people in the team is a high priority.

Each **team member** will know exactly what they have to do and exactly when to do it. If any get it wrong, the race could be lost. So each member will know their own roles and **responsibilities**.

A pit-lane team working in high stress conditions to change the wheels and re-fuel the car

Types of team member

A team is made of different important members.

* Shaper – often enjoys the challenge and has drive to overcome problems, seeks to impose some shape.

* Implementer – turns ideas into reality.

* Completer/finisher – likes to complete tasks to the last detail and on time.

* Co-ordinator/chair – controls the way the team works, identifies what needs doing.

* Team worker – co-operative and diplomatic, avoids friction between members.

* Resource investigator – explores and reports back on ideas from outside the team.

* Company worker – plans practical ways forward and agrees procedures.

* Monitor/evaluator – monitors a situation, looks at facts and makes accurate judgements.

Activity

We all think we know how to make a cup of tea, but you would be surprised how many different opinions and ways people can come up with. Work in a group of four people and decide on the best way to make a cup of tea. You will need to make a detailed list of what needs doing and in what order. After you have finished, make a further list of what went well and what did not in trying to function as a team.

Activity

Have a look at the list of different types of team members from shaper to monitor/evaluator. There are several recognised ways of gaining an insight into the type of team player you may be. The Belbin test is probably the most widely used method. Either by searching for this on the internet, or by asking your teacher to give you instructions for this test, find out what type or types of team member you are more likely to be. Does the outcome of this test meet with what you expected?

Just checking

* Do you know what you will be doing throughout this unit?
* Are you convinced that everything in our world has something to do with engineering?
* What stages does a team go through before it operates efficiently?
* What type of team member are you?

Communications within teams

Most of us have heard of the game 'Chinese whispers'. Often we think people have understood what we meant when in fact they haven't. Outcomes and decisions from team work must be made clear to everyone involved. Correct records will form the basis of what gets done and what does not. Understanding each and every team member's roles and responsibilities is vital, if the overall wishes of the team are to be carried out. In this topic we look at effective communications that support each member, and the team as a whole.

The two sides of communications within teams

Communicating outcomes is vital for everyone to be 'rowing the same boat' and working towards the common goals agreed. The second, but less obvious side, is that it facilitates and promotes creativity, motivation and discussion, just to name a few. This can be during team meetings or teamwork activities. It includes being able to ask the right question, and responding in certain ways.

Everyone is present at the briefing but does anyone know what is happening? At least one member of the team is making notes. If these become minutes and a list of outcomes and actions, perhaps the other members of the team will be able to read this record and understand what happened and what is required.

Team meetings

Some meetings achieve vast amounts of progress in a short time while others ramble on and achieve very little. For the meeting to be a success control throughout the meeting is required and a strong chairperson is required.

Team meetings have several aspects:

✳ **Is the meeting appropriate?** – If you need to share ideas, communicate quickly, have interactive reasoning and the ability to solve problems that span far more than the knowledge-base of one member, then a meeting would be appropriate. In an industrial setting, a team would have 'terms of reference', which are formal reasons for the team to exist and lays out what and how things will be covered.

✳ **Someone must be in charge** – A **chairperson** needs to be appointed to ensure a compromise between having a healthy debate and moving the meeting forward to make decisions.

✳ **What is the purpose of the meeting?** – Every meeting should have a purpose. Usually the chairperson sets the purpose and proposes a date, etc. A production team, for example, may meet once a day, once a week or once a month to discuss and solve newly found problems.

✳ **An agenda is required** – It should be circulated in advance to ensure team members are forewarned about what will be discussed.

✳ **Rules for conducting the meeting** – In some meetings, tensions might be high and some members may be in conflict with others. It is appropriate to have ground rules such as: only one person may speak at a time; all have equal rights to speak; the chairperson's decision is final, etc.

A team briefing around a machine – but does everyone know what is going on?

Chairperson – the person appointed to lead a meeting, who can steer the team

Open question – a type of question that will lead to debate and discussion

Closed question – when a factual type of answer is required

Body language – communication without words (e.g. smiles, movements, eye contact, etc.)

	CNC Manufacturing Team		
	Meeting to be held on Wednesday 10th October 2007 at 5 pm in the Training Centre		
	AGENDA		
1	Apologies for absence	2 minutes	I
2	Acceptance of Minutes of meeting held on 17th September 2007, copy enclosed Matters arising	5 minutes	R
3	Production targets CNC Machining update Support functions	20 minutes	A
4	Health and Safety issues Report by Mr R Righton	10 minutes	I A
10	Any other business		
11	Date of next meeting 26th November 2007 in the Traing Centre		I
	Classification: I = Information/discussion. A = Action. R = Ratification		

Outline agenda for a production meeting

✳ **Taking minutes** – This is a record of the outcomes and decisions taken at the meeting. For a production team, the minutes should record who is going to do what and by when.

Notice that the agenda helps those attending to prepare themselves.

Communicating well in meetings

Contributing to meetings is a skill. Some people do this better that others. It is useful to ask '**closed questions**' requiring a specific answer, if a factual type of answer is required, and '**open questions**' when debate and reasoning are needed. It is important to show when you agree or disagree, using '**body language**' as well as spoken responses.

Dealing with conflict

Dealing with conflict brings rewards for a team. Again, some people are naturally better than others but it is a skill that can be developed. These are some of the ways to help deal with conflict.

✳ **Avoid** – This means walking away from arguments, but may not always be an option in a production team. However, you may remain quiet while others resolve the issue – but then must accept the outcome.

✳ **Accommodate** – Often there are solutions that can accommodate both sides in an argument.

✳ **Compromise** – With sensible team members, this is often the best outcome.

✳ **Collaborate** – Team members should feel able to speak their mind, be listened to, and feel they are a critical part of the solution. So each member should respect and listen to others, try to understand their point of view, and actively work towards a mutual decision.

Activity

Prepare a template for a meeting agenda then issue an agenda using your template. The agenda is for a production team progress meeting for manufacturing chess set pieces, each piece machined on a CNC lathe. The supply of material is late and the machines are working at full capacity. Use a word-processing package to develop the template and write the agenda.

Activity

Look at the outline agenda for a production meeting. You are to attend the team meeting as a team member. Make notes on the agenda against each item: what you already know, any fears you have and any questions you want to ask. Share this with your group and exchange notes.

Functional Skills

In a group of three, practise using closed questions. Ask questions like: 'How old are you?' 'How many holidays do you have each year?' Discuss how this contributes to your discussion.

Now practise using open questions. Ask questions like 'Do you think you are the oldest in your class/group?' 'Would you like to have more holidays each year?' Now discuss how this contributes.

Just checking

✳ What should be sent out in advance to prepare for a meeting?

✳ What formal record is made of decisions from a production meeting and what should it record?

Production information

Production information is exactly what it says – information about how a product is made or produced. For this unit, we really mean all of the information needed to plan and make the product. It will normally be in the form of engineering drawings, sometimes supported by a product specification and other documentation – sometimes called technical drawings. Different types of drawing are covered in Unit 4. Some, such as an assembly drawing, show how a product fits together; others, such as an orthographic projection, give details like **sizes** and tolerances, to plan how the product will be made.

The information

The specification and drawings, and other documents about a product, should provide information that cannot be misinterpreted. This could include: **size**, **shape or form**, **materials** to use, **quantity**, **process methods**, **quality**, etc.

Types of drawings as part of your production information

You may have already come across the range of drawings we use in engineering, but it is important just to remind ourselves what they are.

* Freehand sketches
* Isometric and oblique projection
* Block and flow diagrams
* Schematic and circuit diagrams
* First and third angle orthographic projection
* Assembly and exploded diagrams

One of the most basic things you need from a drawing are sizes (we call them dimensions). The first drawing (left) is a third angle drawing of a simple round component, showing the length and diameters of its features. The second is an isometric drawing showing a rather more complicated block, again with its features dimensioned.

The trolley jack is designed to lift a car up. Think of all the information that would be required to plan how to make it. We can see what it looks like, but what are its overall dimensions? What materials is it constructed from? How many hexagonal nuts are required for each trolley jack? The questions go on.

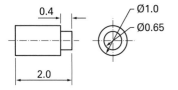

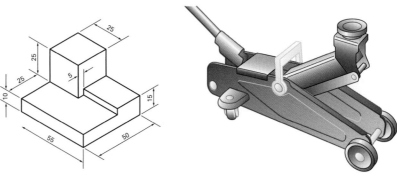

A third angle and isometric drawing

A trolley jack often used in car repair workshops

A breakdown of the information required for production

Six categories or pieces of information relating to production were listed above, which are found in product specifications, drawings and other documentation.

✳ **Size** – Many products have become smaller due to improvements, for example, in materials technology. As well as overall sizes of the product, the size and positions of features such as holes, slots, angles, etc. will be included. These are shown as dimensions from a definable reference point.

✳ **Shape or form** – Often the shape is determined by where or how the product needs to be used or positioned. Improvements can be made to the strength characteristics by selecting certain shapes. The designer will have made sure that the product can be manufactured or assembled, ensuring that thicknesses of sections or parts of the product give the required strength characteristics, but without adding extra weight or waste. Sometimes parts need to be welded together, or cast as a single item, to achieve these requirements. Weaknesses in corners can be avoided, for example, by designing simple radii into the corners.

✳ **Materials** – The designer will have carried out development work and calculations to ensure the materials chosen are strong enough to withstand the stresses the product will be subjected to during its use.

✳ **Quantity** – Sometimes 'one off' products are required, such as a replacement special bearing for a machine; other products will need to be made in large quantities. Often a parts list will show this.

✳ **Process methods** – Every company has a limited range of processes. This information will be about the processes available and their capacity.

✳ **Quality** – This will specify the accuracy required in manufacture. It is impossible for products to be made to an exact size, so the designer will specify a variation (tolerance) on the exact size that will still allow the product to be assembled and to function properly. Another typical feature might be surface texture which, if important, will be specified.

Size – the dimensions of each part, position of features, etc.

Shape or form – the overall three-dimensional representation and what the product looks like

Materials – what material the product is made from

Quantity – how many of each product and parts of the product to make

Process methods – how it will be made

Quality – tolerances in dimensions and other essential features

Functional Skills

For one of the drawings shown, work out the volume of the material for the given component. The dimensions on the simple round component are in centimetres and, for the block, the dimensions are in millimetres. While the round component may have fewer calculations, you will need to find out the formulae to use. When finding the volume of material used within the block, you will have to break the work down into smaller tasks and visualise in detail what the drawing is showing. You may ask your teacher to help you visualise this. If you feel confident, you may wish to work out the volume for the components shown in both drawings.

Activity

For a product you are familiar with, think about and make a list of the service requirements, economic requirements and manufacturing requirements, which could be split into material and product requirements. You may have to ask your teacher to give you more information about these requirements. Having done this, try to put them into the different categories of information.

Just checking

✳ What information will you need to plan and then make a product?

✳ What sources will give you this information?

Dimension – a measurement in respect to the size or feature on an engineered product

Tolerance – an allowable deviation from the desired size (no size can be achieved exactly)

A production engineer at work

Interpreting engineering drawings for planning details

In the last topic, we looked at the range of information we need to plan for, and then make an engineered product. The function of extracting information from the drawings, product specification and other documentation is usually carried out by a production or planning engineer. In some engineering companies, this activity will be carried out at the same time as the design is developed. This unit is mainly concerned with planning for manufacture by CNC machines.

Recognising the information

For part of this unit, you will be working in teams to extract information from engineering documentation. The unit specification states that you must produce evidence for all the criteria on an individual basis, so it is important that, as an individual, you can interpret the information given on the drawings and product specification. You will be asked to describe this information in your assessment. The process is something that would be done at the planning and manufacturing stage when making a quantity of an engineered product using CNC equipment.

Production or planning engineers have a wide range of experience that enables them to look at the information supplied by the design team and interpret this into instructions for the CNC machine setter and operator. This is part of planning the manufacture.

Working with the information

Let's look at each category of information and see what we may have to do for a planning activity to be carried out.

❋ **Size** – Look for a reference point. Are the dimensions all in mm? Do we need to add **dimensions** together or subtract one from another to get overall sizes? What size are the holes and where are they (if any)? These sizes will enable the production engineer to plan 'profiles to be machined'.

❋ **Shape or form** – Although it may appear that this information is not that important, without it we are unable to decide what process or processes need to be used. A one-piece complex shape is most likely to be cast or moulded. However, CNC machines can produce complex shapes in three dimensions. The shape or form helps us decide 'cutter paths'.

❋ **Materials** – The specification should state the material to be used. The production engineer needs to align this information to sizes of material stock available. For example, the finished size of a shaft may be 22.5 mm diameter, so the nearest material size above may be 25 mm diameter, allowing for machining. We also need to consider how hard or easy that particular material is to machine.

❋ **Quantity** – This is probably the least important concern when planning for CNC machining but very important for scheduling. CNC machines are efficient for producing 'one-off' and 'batch' quantities.

❋ **Process methods** – Are there processes available to suit the requirement in terms of size, accuracy, speed, etc? A 210 mm diameter work piece will not go into a 150 mm maximum diameter chuck on a CNC lathe.

❋ **Quality** – This will enable the production engineer to decide where and when to take measurements during and after manufacture. Knowing the

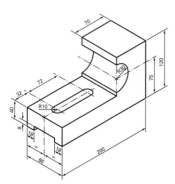

An adjustable mounting block

tolerances set helps them to decide which manufacturing processes may be suitable – and, when considered with the quantity required, the inspection or 'gauging' techniques to use. Often drawings have a general tolerance and you need to know where this is shown on the drawing.

The adjustable mounting block is well specified from the size and dimensions point of view. It is also possible to see the shape and form, but many of the other categories of information are missing.

Preparing a plan for production

Production planning is not just about the sequence of operations: it involves the selection of components, materials, tools and equipment, application of health and safety and quality checking. The following is a tick list for preparing a production plan.

✳ Do you have the specification, drawings, etc.?

✳ Use the parts list, or draw one up yourself, and decide whether each part of the product will be made or bought in from suitable suppliers.

✳ Decide how to make the product. What processes are involved?

✳ List the tools and equipment needed for each part of the product manufacture/process. Work out the speeds and feeds required by each.

✳ Work out the sequence of manufacture (the operations).

✳ Use the tolerance information to decide on quality requirements.

✳ State the health and safety requirements for each operation (PPE, guarding, etc.)

How complicated is planning?

By now you should have discovered that planning to make an engineered product is not a simple activity. To decide what processes are required, you need to know several things such as sizes, shape, quality and tolerances. Once the process is chosen, setting it up will also be influenced by other information found in the product specification and drawings, such as type of material. Some materials need a slow 'speed and feed rate', while others need a fast rate. Some will need a 'roughing cut', while others will not.

Production planning

In this topic, you will look at the actual activity of production planning. The product specification, including drawings and other documentation, will be forwarded by the design team to the production planning team. The way manufacture is planned can make or break whether production is efficient. Poor planning leads to poor utilisation of staff, machines and other resources. Once a production plan is developed, a schedule will be prepared to take account of the quantity required.

What is in a production plan?

Although there will be differences between engineering companies, the following items would be found in a typical production plan:

* sequence of operations
* materials
* process methods
* tools, equipment and machinery to be used
* critical production and quality control points, inspection and quality checks
* health and safety requirements.

What will a plan look like?

Some of the requirements in a production plan will be found within a 'title section' and others within a 'table section'. Here is a layout of a typical production plan.

Reviewing what you need

Think back to what you did in the topics on production information and interpreting engineering drawings. To start production planning, you need to be very clear on what you need to have and what you need to do. Make up a checklist of the information you would need from a product specification and drawings.

Cutting speed – usually quoted in metres per minute, but the machine has to be set at revs per minute, so be careful when using this terminology

Feed rate – the rate at which the tool moves across or into the material as it is taking a cut, quoted in metres per minute or, more likely, mm per rev

Drawing number: W250-19919	Component name: Gearbox small shaft		Notes: Issue 1	Sheet 1 of 07
Material: Bright drawn mild steel 35mm dia × 230mm long	Planners name: Dan Evans		Checked by: MD	Date: 3rd July 2008

Seq	Process	Tools & equipment	Speeds & feeds	H&S	Control points	Inspection & quality	Time (minutes)
10	Check material and deburr edges	Vernier calliper File	Hand	Gloves		Size 35 mm dia × 230 mm No rough edges	5
20	Turn first side Mount billet Turn 25 dia x 55 Turn 20 dia x 35 Undercut and chamfer	Denford 125mm Easiturn CNC Lathe 3 jaw chuck Knife tool Undercut and chamfering tool	290 rpm 0.2 mm/rev 120 rpm 0.05 mm/rev	Guard in place Eye protection	Check tolerances on 25 and 20 dia Check undercut	External micrometers External spring callipers	11

Using an engineer's tables to work out speeds and feeds

A CNC milling machine

When you work out speeds and feeds, you need to either find this information directly from 'engineers tables' or use formulae to work them out. The main factors are the type of material being cut, the tool being used and the type of finish and tolerance required. Aluminium alloy is cut at a much faster rate than mild steel, for example.

How important is the information in the production plan?

Let us look at some of the features of the production plan.

* If we get the order of the processes wrong, a subsequent process may be impossible to carry out.

* The way the product is held on CNC machines is very important.

* Although the material will be specified by the designer, the planner has to work out such things as billet length to ensure the machining features can be achieved, the work piece is secured and there is minimal waste.

* The selection of the correct tooling is vital, if the outcomes and features required by the drawing and product specification are to be met. Some tools are designed for taking large cuts and some for smaller 'finishing cuts'. Some work piece shapes are dependent on the shape of the tool and some are not.

* If the wrong speed or feed is used, tooling could be damaged or burnt out. The surface finish obtained is dependent on the tool shape, speed and feed rate.

* If quality checks were not specified, products could be made that do not conform to requirements and further processing would be a waste of resources and time.

* Probably most important, the health and safety requirements should be specified. These must be correct, so that any operator following the plan is working in a safe and approved manner.

If production and planning engineers have done their job correctly, manufacture of the engineered product will go smoothly. CNC machines are expensive to buy and rely on accurate planning to ensure they are utilised well.

Scheduling for production

Production scheduling is a very different activity to production planning. In many engineering organisations, it will be carried out by a different department. The design solution, in the form of a product specification and set of drawings, is handed over from the design office to the production engineering department. Generally, a production engineer is responsible for production planning, whereas a production controller is responsible for production scheduling, taking into account the production plan, sales forecast or known demand.

What do we need to consider for production scheduling?

Once a production plan for an engineered product has been completed, a timescale for the manufacture of a quantity of it is required. The following need to be considered:

* the time required for each operation
* the sequence of manufacture
* whether operations can be done 'concurrently' (at the same time)
* the completion date
* whether there is a reliance on outside sources or supply
* the amount of capacity available on the machines and resources required
* if you are scheduling for a team manufacturing activity, you will also need to consider the skills and preference of each member of the team

Remember: a schedule is a representation of the time taken to carry out a task – for this unit, the manufacture of a quantity of an engineered product.

A **Gantt chart** shows the time taken to carry out an activity or operation, represented as a coloured block on a time series. In the example, the different colours show the different stages of the operations or activities. Such charts can be produced as a table in a word-processing package, a spreadsheet package, with cells to represent a block of time or activities, or with specific scheduling software such as Microsoft Project™.

Aspects of scheduling

The scheduling approach will largely depend on the production quantity required, which depends on what is intended to happen to the product once manufactured. There are costs associated with holding stock and setting up for further production. If quantities are to be held in stock and released to customers over a period of time, **batch production** is used. We can use a formula to calculate an 'economic batch quantity' to minimise cost.

The term 'batch production' can also be used when a client orders a given quantity to be delivered at a certain time. In this case, the schedule can be worked backwards from the date the customer needs the

In what order should items be installed?

Have a look around the room you are in. List all its features, such as the electrics, seating, decoration, etc. Now try to put them in the order that they needed to be installed. Alternatively you could consider a car, and try to list the order for assembling the different parts. Consider whether any could be carried out at the same time (we call this **concurrently**).

Concurrent operations – more than one activity carried out at the same time

Batch production – a quantity of engineered products manufactured in one run, that will need to be repeated if further quantities are required

Gantt chart – a representation of a schedule of operations or activities along a time series

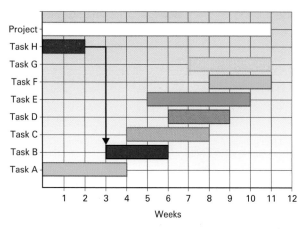

A part schedule shown as a Gantt chart

A machine that appears to be in a poorly scheduled system

goods to provide a start time for manufacture. Some 'safety time' may be built into the schedule.

Scheduling for high-volume production is usually dominated by the capacity of the equipment needed. Resources may be temporarily dedicated to one product. In this case, balancing each part of the process is more important than considering what needs loading onto what machine at what time (as when scheduling batch quantity 'runs').

There are different ways of presenting schedules. They include Gantt chart, bar chart, flow chart and network diagrams.

Think of the waste shown in the picture, with all of the work that has gone into making the parts that are stacked up awaiting processing. Has the customer been satisfied by receiving his order yet? No! It is important to schedule production to maximise customers' delivery expectations but not waste resources.

Problems with scheduling

Any manufacturing company must be efficient to become successful. A large contribution to this is the effectiveness of production scheduling based on a sound production plan. Unfortunately, internal pressures are often in conflict and can result in different approaches being adopted.

✳ **Marketing override** – This approach seeks always to meet customer expectations and delivery dates. It is usually achieved by backward scheduling. If extra capacity is required to meet this delivery, this may be accommodated by additional working and overtime, or by sending work to be done outside (subcontracting).

✳ **Production override** – This approach seeks always to maximise the capacity of the resources – human and machine – minimising set ups.

An overview of processes including CNC

A range of processes was addressed in Unit 4. In this topic, you will look further at some of the CNC processes that you are most likely to come into contact with.

So what are typical processes used for?

* **Grinding** – Several types of machine exist for grinding, but those associated with precision manufacture are likely to be surface and cylindrical grinders. Some remove material slowly but have advantages when it comes to machining hard materials, as other processes may not be able to do this. Also they can be accurate to 0.001mm. The surface finish is very smooth.

Milling machine

* **Milling** – This uses cutting tools that usually have multi-points and can remove material relatively quickly. The tool usually rotates and the work moves relative to the tool. Milling machines can produce parallel, perpendicular, inclined plain surfaces, pockets, islands, slots and holes.

* **Turning** – This uses single-point cutting tools to 'generate' or 'form' surfaces in a cylindrical way. The work piece usually rotates and the tool is made to move relative to the work. The tool can traverse along the axis or at 90° to it, or even at some other angle to it.

* **Presswork** – This can involve either 'punching out' holes or shapes, or bending, folding or forming into shapes.

Why work holding is important

To set up processes, you must know the range of work-holding devices available to support items to be worked on.

* **For grinding**: chucks, collets, centres, faceplates, machine vices, magnetic plates and chucks, clamps, angle plates, vee-blocks and magnetic blocks
* **For milling**: clamping direct to a machine table, magnetic table, machine vice, angle plate, vee-block and clamps, and rotary table

What processes do you already know about?

Visit a local engineering company and make a list of the machines and processes you find there. If you cannot do this, visit your school or college engineering workshop and carry out the same exercise. Can you relate any of the items on your list to machines and processes you came across earlier in this unit?

All these processes can be applied in CNC applications, though the more usual applications are turning and milling.

The traditional manually operated milling machine shown and lathes are still found in most engineering companies because of their versatility.

The basic tooling and work piece movements will still be evident on their CNC 'cousins'.

Activity

Find an image of a commonly used cutting tool for either a lathe or milling machine. Copy this into a word-processed document with empty space around all sides. Print copies and make a sketch showing the tool as if it was making a cut. Show your teacher what you have done.

* **For turning**: chucks with hard jaws, chucks with soft jaws, collet chucks, drive plate and centres, fixtures, faceplates, fixed or travelling steadies and four jaw chucks

* **For presswork**: clamping direct to a machine table, feed mechanisms, fixtures and blank holder.

Tooling is important when machining products

You also need to know the range of tooling that can support processes.

* **For grinding**: grinding wheels such as cup, flaring cup, straight sided, recessed or double recessed wheel, dish, saucer, disc and segmented

* **For milling**: face mills, slab mills/cylindrical cutters, side and face cutters, slotting cutters, slitting saws, twist drills, end mills, slot drills

* **For turning**: turning tools, facing tools, form tools, parting off tools, thread chaser, single point threading, boring bars, recessing tools, centre drills, twist/core drills, solid reamers, expanding reamers, taps, dies, knurling tool

* **For presswork**: punch, bolster, die, stripper.

Each tool is used to meet certain requirements. The motions of the tool in relation to that of the work piece will either 'generate' a shape or 'form' a shape. We call this the 'kinematics of machine tools'.

So what is different when machining products using CNC machines?

By definition, CNC machines involve the use of a computer to control a range of machining functions, such as cutter path, speed or application of cutting fluid. A '**program**' may be generated several ways. In many manufacturing systems, it may be downloaded straight to the machine from information stored in a computer-aided design (CAD) package, or via tapes, CDs, etc. It can also be entered manually at the CNC machine console.

The CNC machine enables the programme to have a 'dry run', simulating the process without cutting material. This is useful to 'prove' correct motion and actions within the program. Often the programme will contain 'loops' of information that define activities and actions. These enable a repeat cycle to be set up, such as: tool move fast to work, feed while cutting, dwell, fast tool retract and fast tool return – a 'standard cycle' when drilling holes.

All tool movements and cuts are referenced from a '**datum point**'. This reduces the need for the part to be held in a **fixture** as with manual machining techniques. Tooling can be changed manually or automatically. Machines also have facilities to rotate tools using a tool post similar to facilities on a manual machine.

Activity

Researching work – holding devices

In a team of four, decide which of you is going to research which process (grinding, milling, turning and presswork). Then use the internet and capture images of some of the work-holding devices mentioned. Save these into a Microsoft Word document and print a copy for each member of the team.

Milling – material removal process used to produce surfaces and shapes with the tool normally rotating

Turning – material removal process typically generating/ forming shapes cylindrically with the work rotating

Work-holding – device or devices that position work to meet the required cutting motion

CNC program – a set of instructions that the CNC machine will convert into actions

Datum point – a point that work and tooling movements are referenced to

Fixture – a work-holding device that positions the work piece in the same place each time

Just checking

* What processes may be used when manufacturing a quantity of the same simple product?
* What work-holding and tooling requirements may be used for such processes?

CNC turning

The use of a lathe, when carrying out turning operations, is one of the most fascinating things to watch in an engineering workshop. When this is done with the help and advantages of new technology, it creates even more interest. Once the machine has been set up properly and checked for correct performance and output, it 'blindly' follows its instructions one after another, until the batch quantity required has been completed. Some modern machines can even raise alerts to the operator's mobile phone when a problem arises.

Available designs and models

Some of the most widely used brands and models of control used on CNC lathes are Fanuc, Haas, Mazak and Okuma. Each has its own advantages and disadvantages. All CNC lathes are likely to have a computer console for data entry, editing, etc. Some machines have what might be described as a 'traditional tool post type' tool holder, while others may have a turret tool holder. Many provide automatic change over of tools.

The turret housing the tooling is very important, as this allows tools to be positioned accurately. When programmed relative to the datum point, this will produce accurate components and features. The console on the machine is in a comfortable position for the operator to use.

Machining considerations

For tooling movement to be in the desired way, a relationship must be established between the machine and work piece and is provided by a datum point. A **work piece datum** may be defined as a point, line or surface from which all dimensions are referenced. The production engineer will make a sensible choice about this during planning. Once the work piece is on the machine, a relationship between this and machine movement needs to be established, using a machine datum. The tool movement in most lathe work only happens in one of two directions (the 'Z' and 'Y' **axes**) or a combination of these. The 'Z' axis is in line with the work piece axis. It is convenient to locate the 'Z' zero datum at the end of the workpiece furthest from the chuck or work holder. The 'X' axis is at 90^0 to the 'Z' axis. The zero datum in this axis is conveniently located at the centre line of the component.

Tooling requirements

In the last topic, we discussed tooling movements as either **generating** or **forming** shapes on a workpiece. The photo shows a range of typical tools used on a CNC lathe.

The boring tool, facing tool, and knife tool are typical tools that generate the required shape or surface. The radius tool and chamfering tool form the required shape or surface. The knurling tool actually deforms the surface of the material. Is this generating or forming? Both the undercutting and parting off tools rely on generating principles, but the size of the 'cut' is dependent on the tool shape/size and, therefore, also relies on forming principles. In essence then, generating a shape does not rely on the tool shape, just the tool and work movements; the forming

---

How are CNC lathes specified?

In a group of three, each use the internet to search for information about three different CNC lathes. Obtain the specification and any other information you can find and present this in the form of a brief word-processed fact sheet. Then get together and discuss which one you wish to 'adopt' as your group's findings. Show your teacher so the findings can be distributed to the others on your course.

Work piece datum
– a reference point that measurements are taken from, which may be negative

Axes – the directions the tool is able to move in, usually by reference to three axes at right angles: X, Y and Z

Generating shapes – relies on tool and work movement to produce a shape

Forming shapes – relies on the tool shape to produce a feature or shape on the work piece

of shapes relies on the shape of the tool. However, what would you say about the 'rounded' facing/turning tool?

Tools can be made from a range of materials, including high-speed steel and tungsten carbide.

Tool offsets and tool data entry

Working out and setting tool offsets, and entering tool data, are skilful activities, which your teacher should demonstrate. These include geometry, wear, tool nose radius and tool orientation. Data screens or consoles are used for tool offset entry. Tool data is obtained from standard tool catalogues and machining handbooks. Often quality and size problems during use are overcome by adjusting offsets.

Data entry is made easy as some machines have a visual display to ensure the operator can see what is happening as data is entered.

Typical tools on a CNC lathe

Tooling can be mounted on a turret for ease of access

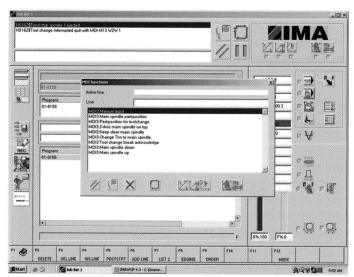

A typical display on a console or computer screen of a CNC lathe

Activity

Visit an engineering company with CNC lathes and talk to the operators. If this is not possible, investigate a CNC lathe in your school or college workshop. Find out how one is programmed. Write a brief report describing what you have seen. Explain how it works and illustrate your report with relevant sketches, diagrams and photographs. Share what you have done with your teacher and others studying with you.

Functional Skills

Find the formula used to work out the revs per minute (RPM) to set on a CNC lathe when machining a shaft. This is something you should have done in Topic 5. Use the formula to calculate the RPM for a 45 mm diameter shaft made from mild steel, being cut using an HSS knife-turning tool. You will have to find the correct cutting speed to use for these conditions. What would happen to the machine RPM setting if the material was changed to an aluminium alloy? You should back up your statement with reasons and calculations.

Just checking

* How are CNC lathes specified?
* Where are the best positions for work piece datums?
* What types of tooling generate or form shapes when carrying out a turning operation?
* What are tool offsets when setting and operating CNC lathes?

CNC milling

Like a lathe, a milling machine is also fascinating to watch in an engineering workshop. Try to see a CNC milling machine working in an industrial environment. Some modern machines have automated quality checking, work feed mechanisms to position the work piece before and after machining, and swarf conveyors. Often the work piece is completed at one setting.

Available designs and models

CNC milling machines are often referred to as a machining centre. This is because of their ability to produce complex 3-D shapes that traditional milling machines cannot do. Some of the most widely used brands and models of control used on CNC milling machines are Fanuc, Fazor, Haas, Mazak, Heidenhain and Okuma. Each has its own advantages and disadvantages. All CNC milling machines are likely to have a computer console for data entry, editing, etc. The machine table always moves in the 'X' and 'Y' axes and most also move in the 'Z' axis to create the tooling movement up and down. These are **3-axes machines**. Others also rotate around these axes, making 4th and 5th axes of movement.

The machine shown on the next page is a 3-axes machine. Like CNC lathes, the console on the machine is in a comfortable position for the operator to use. CNC milling machines can exist in virtually any of the forms of their manual counterparts. The most advanced CNC milling machines have five axes. Horizontal milling machines have a 4th axis ('C' or 'Q'), allowing the horizontally mounted work piece to be rotated. The 5th axis ('B') controls the tilt of the tool itself. Extremely complicated geometrical shapes, such as that of a human head, can be made with relative ease with such machines. The skill to produce such programmes is beyond most humans. Therefore, 5-axis milling machines are practically always programmed with computer aided manufacturing (CAM) software.

Machining considerations

As for turning, for tooling movement on CNC machines to do what is required, there must be a relationship established between the machine and work piece. A work piece zero datum is used from which all dimensions are referenced. This is often the top surface and bottom left hand corner of the work piece, providing a datum in three axes (X, Y and Z). In some cases, it is easier to consider this datum to be some distance away from the work piece. The production engineer will make a sensible choice about this during planning. The machine datum will usually be given by the manufacturer and is often the bottom left hand corner of the table. To obtain this relationship between machine and work piece a fixture will often be used to locate the work piece. This, in turn, can lead to automatic work piece placement.

How are CNC milling machines specified?

In a group of three, preferably different people from the last topic, each of you use the internet to find information on a different CNC milling machine. Obtain the specification and any other information and present this as a brief word-processed fact sheet. Then agree which of the three you wish to adopt as your group's findings. Show this to your teacher so they can distribute the findings to the others on your course.

3-axes machines – CNC milling machines with relative tool and work piece movement in three directions (X, Y and Z axes)

Quick-release disposable tips – usually made from carbides or other ceramics and often golden in colour, can be supplied in different shapes and can have more than one cutting edge to allow them to be rotated when needed. Removal and replacement is easy and quick

Tool offsets – when tools are mounted, the amount of extension of different tools from the tool holder varies according to their size and length. Tool offsets are, in general terms, these differences

Typical CNC milling machine

Close up of a carbide tipped face mill

Tooling requirements

Unlike with manual milling machines, due to the ability to interchange tooling, the tooling used is more standardised. CNC milling machines nearly always use SK (or ISO), CAT, BT or HSK tooling. SK tooling is the most common in Europe. CAT tooling is sometimes called V-Flange tooling. It is not too important at this stage that you know about each of these specifically. Two other considerations to make though are the taper type and size, and the thread type and size that allow the tool holder to be mounted in the machine tool spindle.

The tools themselves can be made from a range of materials, including high-speed steel and tungsten carbide. With the high removal rates achieved with CNC milling, it is normal to use carbide- or diamond-tipped tooling.

With the variety of tool movement that can be programmed into a CNC milling machine, complex 3-D shapes and contours can be achieved. The use of **quick-release disposal tips** on the cutting tool enables accuracy to be maintained during tool changes. The operator can change the tips with minimal interference.

Tool offsets and tool data entry

As with CNC lathe tool offsets, working out and setting tool offsets on CNC milling machines, and entering tool data, are skilful activities, which your teacher should demonstrate. They include geometry, wear, tool radius and tool orientation. Data screens or consoles are used for tool offset entry. Tool data is obtained from standard tool catalogues and machining handbooks, but will depend on whether preset or non-preset tools are to be used. Often quality and size issues during use are overcome by adjusting offsets.

Finding out about different types of materials

In a group of three, one of you take metals, another alloys and the third polymers and produces a word-processed list of different types for each. When you have all completed your lists, cut and paste them onto each other's to produce a complete list covering all three types.

Processing materials and safe working

It is the job of the designer to specify what material a product is to be made from. They will have referred to material properties and carried out calculations to ensure the correct type and size of material is used so that the product will not break or fail during use. Often this leads to problems when machining the product. Although designers have wide experience of machining techniques and their limitations, sometimes they have to specify a material that will cause difficulties. Here you will look at some of these difficulties, along with working safely when processing materials.

Conflicts between the requirements of the finished product and machining

Two simple examples demonstrate this conflict, which sometimes results in a compromise.

* **Hardness** – The product may need to be made from a material that is hard and resists wear, but during manufacturing this is one of the limiting factors on how a product can be machined. Hard materials are difficult and may be slow to machine.

* **Smooth surface finish** – The product may need to have a very smooth surface, perhaps to allow smooth movement of one surface over another, but not all processes can produce a smooth finish on certain materials. Some aluminium alloys can be machined very quickly and with heavy cuts while leaving a smooth surface. The same processing conditions applied to high-carbon steel would not necessarily produce the same smoothness on the surface. In this case, perhaps a slower cutting rate or smaller cuts would be required to get the same desired smoothness.

Machinability of materials

When we machine a work piece we produce swarf – the excess material removed to leave the desired shape and features. The way swarf is generated is called **chip formation**. Machinability can be defined as the relative ease with which the chip may be separated from the material being machined. The two main aspects are:

* the properties of the materials being machined
* the machining conditions involved.

In activities in previous topics, you used a cutting speed obtained from engineering handbooks and other information. This is a very complex subject, but this cutting speed is basically the best speed at which to cut the given material to give the least overall cost. It considers, along with the work piece material, the life and material of the cutting tool.

Either a 'continuous chip' or a 'discontinuous chip' indicates successful cutting conditions for different materials. The continuous chip sometimes causes swarf disposal problems and often a '**chip breaker**' is deployed as part of the tooling set up. When selecting the correct tooling for any CNC work, you will need to consider whether to use chip breakers or not.

A CNC turning operation and a close up of the swarf (chips) produced

Machinability – how easily a chip can be formed and separated

Chip formation – how swarf is generated from a cutting action between tool and work piece

Chip breaker – an aid to control the flow of swarf-chips from a cutting action in a continuous flow

The following table represents how easy some typical materials are to cut:

Material	Machinability
Stainless steel	Difficult
Thermosetting plastics	Fair
Alloy steel	Fair
Thermoplastics	Fair/good
Mild steel	Fair/good
Cast iron	Good
Aluminium alloy	Good/excellent
Brass	Excellent

This is only a simple guide, as many other aspects come into play. For example, thermoplastics require a lot of support by the work holding device to maintain dimensional control, and careful application of cutting fluid.

Working safely

In the UK, workers are protected from working in hazardous ways and also have responsibilities to ensure that what they do is in a safe manner, protecting themselves and others. All engineering companies need to follow the Health and Safety at Work Act 1974. When you are working in a team to produce a quantity of a simple engineered product, you will be following a production plan that you prepared earlier. If, at that point, you did not get this checked by your teacher, you should ensure you do so before you use it, as it should contain important information about health and safety. It will highlight the need to have guards in place and what personal protective equipment (PPE) to use.

Each person working in an engineering environment has the responsibility to prepare themselves to work safely. Although not always specified in the production plan, some of this PPE is expected as a matter of course: for example, overalls and safety shoes should be worn correctly at all times when working in an engineering workshop. Before you work on any machine, you should also ensure you are familiar with all the safety features of that machine, including any emergency stop buttons and guard positions and other requirements.

Functional Skills

Many tooling manufacturers have produced internet videos showing what their tooling can achieve. On your own, find some of these video clips and make a note of the website address. Try to find applications of machining different materials. Share what you have found with the rest of your course.

Activity

As well as wearing the correct PPE and following safety requirements while working a machine, you need to be aware of what to do if things go wrong. Find out about the requirements of the following four regulations: Control of Substances Hazardous to Health (COSHH) Regulations 2002; Manual Handling Operations Regulations 1992; Reporting of Injuries, Diseases and Dangerous Occurrences Regulations (RIDDOR); Health and Safety (First Aid) Regulations 1981.

Just checking

* Why is it important to know about the type of material being cut?
* Why is it important to apply the correct speed for given cutting conditions?
* What PPE should you wear when working on machines?

What evidence of different accuracies can you find around you?

One of the main reasons for applying tolerances to the dimensions of a work piece is to ensure that, in applications, different 'fits' can be obtained. Find out what is meant by 'clearance fit', 'interference fit' and 'transition fit'. Once you have done this, visit the school or college workshop and see if you can find where these different 'fits' would be applied in some of the engineering products you find around. The best places to look may be within the design of the machines there.

In the last topic, you saw that it is the job of designers to specify what material a product needs to be made from and that this can sometimes conflict with ease of machining the product. Here, you will look further at some of these difficulties and, in particular, how quality is checked when producing work pieces.

Conflict between accuracy and tolerance requirements of the finished product and machining

No product can be made to an exact size, so we have a series of tolerances that can be specified to meet given conditions for a specific design solution. Sometimes this leads to the production engineer having to think very carefully about which process to use when planning for production. Two simple examples demonstrate this conflict, which sometimes results in a compromise.

* **Accuracy** – Often there is a need to specify a high level of accuracy, which would be in the form of a tight tolerance. Without this, for example, two interconnecting parts may not fit together. To obtain this accuracy, there may be a need to plan an extra process. After turning, grinding may be needed to ensure the level of tolerance and accuracy can be met. Generally, CNC machines can produce more accurate results than manual ones.

* **Smooth surface finish** – Not all processes can produce a smooth finish on all materials. We call this feature, which needs to be measured, '**surface texture**' to avoid confusion with a 'finish' that is applied to a surface, such as painting or coating. If a high degree of surface texture is required, grinding is likely to be used in preference to milling or turning, which might mean an extra process to get the same desired smoothness.

In both these cases, achievement of high accuracy and surface texture creates higher production costs. In the last topic, we talked about machinability. Again, those materials with good machinability are likely to produce work pieces more easily, with tighter tolerances and better surface textures.

How do you check the quality of work?

In the last topic for this unit, you will look at sampling quality through statistical methods. In this topic, we need to establish the appropriate techniques and equipment to use to check dimensions, tolerances, fit, finish and performance.

The planning engineer will have specified what type of measuring equipment to use. Vernier callipers and external micrometers are two of the most common found in an engineering workshop to check sizes, dimensions and tolerances.

Typical piece of measuring equipment

Alternatives to measuring

If the need arises, sizes and dimensions can be compared against a maximum and minimum limit. This can be done by the use of **gauges** in the form of 'go' or 'not go' gauges. If the 'go' gauge accepts the dimension and the 'not go' gauge does not reject the work piece, it is said to be within tolerance.

The planning engineer will have specified the type of gauging equipment to use to check whether or not an item has been machined within tolerance. A typical plug gauges, to checks holes, is shown below with a 'go' and 'not go' end. Gap gauges checks external features.

Comparison to a known standard

Use of gauges is considered to be a method of comparison, though it is checking to see if dimensions are within limits and, therefore, satisfactory. Other methods involve setting a dial test indicator (DTI) to zero against a known size and then measuring dimensions with the DTI. Often the known standards will be slip gauges.

Surface texture – those irregularities with spacing that tend to form a pattern or texture on the surface. This texture has roughness and waviness

Gauge – equipment that helps decide if a dimension is within tolerance but without declaring actual size

A set of slip gauges

A DTI

A typical plug gauge

More quality control techniques and interpreting data

It would not be good if products were unfit for purpose due to poor manufacturing control. All of the good work of the designer would be wasted. In the last topic, you looked at some of the techniques and equipment used to check the quality of an engineered product. This informs you if a product is meeting the standards laid down, but not whether the process is performing as expected, or whether it may produce poor standard work pieces in the future.

The benefits of collecting and using data

In your discussion group, you probably talked about whether, if the machine settings were not altered, the next work piece produced would also be out of tolerance. Was the decision too premature and a panic reaction? All processes operate under '**normal distribution** conditions'. If you accurately measured every single work piece for a given feature and the measurement technique was able to show small variations, then you would see that these measurements fell into a 'normal distribution'. This knowledge would enable a sampling technique to be used, which would show whether the process is likely to continue to produce features that are within, or likely to go outside, tolerance. It would show trends that could be acted on. Likewise, analysing this data would enable you to make a judgement on the machine's accuracy capability.

Statistical process control (SPC)

One manufacturing technology technique is SPC. It gives an opportunity to control a process and reduce the chances of producing out-of-tolerance work pieces. A chart is developed that the machine operator uses to plot the outcomes from sample measurements. This has control limits below and above a target or average size of the variable being measured, such as the diameter of a shaft. People get confused between control and capability. If a process is under control, it means that it is stable, if the process is capable, it means that items can be produced within specification. A process can be in control but not capable.

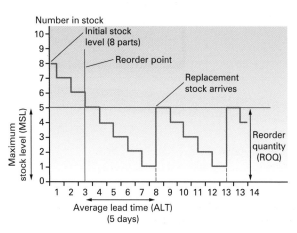

A typical control chart layout

The chart shown is called the '**mean chart**'. A 'range chart' is also needed to give the full picture. This tracks the difference between the largest and smallest measurement within a sample. For a machined work piece, you would be monitoring one of the dimensions, say the diameter. The upper and lower control limits are not the same as the upper and lower tolerances set by the designer. If they were, any work piece outside either limit would be scrapped as not to standard. Instead, these control limits are inside the tolerance limits. If samples measured are shown to be within these limits, the process is under control. If monitored over time, there are several indicators that would indicate that a process was possibly soon going out of control.

What is measured?

So far you have looked at statistical process control as the measurement of features, such as diameters and lengths. This application is known as a '**variables chart**'. In the last topic, you discovered that checks could be made using gauges to decide whether a dimension was in tolerance or not. These can be used as data on a control chart and the same principles applied to controlling the process. However, this application is known as an '**attributes chart**'. The work you do in this unit for your assessment is likely to be with the mean and range charts for variable features.

A plug gauge can be used in an automatic system. Mechanical gauging stations can be located after a machining process, or at the end of the line, to monitor variations in dimensions. The 'go', 'not go' decisions taken by these systems will be automatically fed into a statistical control system that will activate actions on machine settings at the appropriate time.

The requirement when working in your team to produce a quantity of a simple engineered product

The following summarises the actions you are likely to have to take in relation to quality control and statistical methods.

❋ **Taking measurements** – This activity will follow that laid down in your production plan and schedule.

❋ **How you will take measurements** – Again this will follow that laid down in your production plan and is likely to include the use of measuring equipment found in your workshop at school or college, such as micrometers, vernier gauges, plug gauges, gap gauges, etc.

❋ **Plot sample measurements on the control charts** – The control chart will be supplied by your teacher. It may be a computer-generated version and you may have to enter your measurements on the computer.

❋ **Analyse entries on the control chart** – You will need to take a view on what is happening with the accuracy of the work pieces and the performance of the process. Do machine settings need altering, do tools need changing, etc.?

When you enter a point on the chart, you may need to take action depending on whether it falls:

❋ **on or outside a warning limit** – Take a second sample and, if this also falls on or outside the limit, immediately adjust the machine setting.

❋ **outside the action limits** – Assume the process is out of control, stop machining and investigate immediate corrective action.

❋ **within limits but in a series that indicates a trend** – Judgement depends on how many points are within that trend and how close to the limit the last point is. Action is normally required when reaching the fifth point in a row.

Activity

At this level you will be expected to use a control chart to take decisions about whether a process is under control or not, whether settings need altering or whether tools need changing. Here you need to find out how these charts are constructed. Use the internet and books or ask your teacher. You may come across new terminology and more types of control limits, such as 'warning limits'. Make notes to illustrate how a control chart is developed and show your teacher what you have done.

Normal distribution – where the proportion of the measurements taken that lies between the mean and a specified number of deviations away from that mean is always the same

Mean chart – tracks the average size of components over time

Variables chart – measures a characteristic that is variable along a scale, such as a diameter, i.e. something that can be measured

Attributes chart – used when a 'yes' or 'no' decision is being taken, such as a plug gauge, i.e. where a judgement is made rather than a measurement taken

Just checking

❋ Is it better to track what is happening to the accuracy of a feature of a work piece rather than wait until some work pieces are out of tolerance?

❋ Could you use a simple control chart when machining a work piece?

❋ What are you expected to do in relation to using quality control techniques and statistics, when working in a team to produce a quantity of a simple engineered product?

Unit 6 Assessment Guide

In this unit, you will be working as part of a team which manufactures a quantity of the same simple product using modern technology and production methods. The starting point is to carry out a planning activity so that the product will be manufactured in the most effective way using available tools and equipment.

Time Management

Manage your time well as this unit has a number of different components that will have to be researched. Ensure that you keep your work safe and that any work in electronic format has a secure and safe backup.

Be well organised. This is your chance to show that you are an independent enquirer, creative thinker, reflective learner, team worker and self manager and therefore will contribute towards achievement of your Personal learning and thinking skills.

Plan ahead for your work experience and make a list of things that you need to investigate so that you make the most efficient use of your time.

Useful Links

Make good use of your work experience to find out as much as possible about issues that are relevant to your coursework. You should talk to a manufacturing/production engineer about how they will plan, schedule, monitor and quality-control the manufacture of products. Ask them how they operate as part of a team and make sure that you keep notes of what they tell you.

You may need to refer to British Standard BS8888 (drawing standards) when you are extracting information from the drawing of the component/s you are machining. Your teacher will be able to give you a student copy to look at.

Things you might need

Your written evidence should be in the form an A4 process portfolio. Your teacher should give you access to the required software to enable the correct presentation.

A digital camera or mobile phone would be useful so that you can record evidence of practical activities.

To complete assessment focus 6.1, you need to show that you were an effective team member. Make sure that you ask your teacher for a witness statement confirming this.

Remember you will have to carry out workshop activities in order to complete assessment foci 6.3 and 6.4. Make sure that you obtain witness statements from your supervisor and photographs showing you working correctly and safely.

Remember to maintain a focus on team working because it is an essential part of manufacturing technology.

What you must show that you know	Guidance	To gain higher marks
That working in a production team requires co-operation and being able to review how this has worked. *Assessment focus 6.1*	✳ Work as a member of a production team. ✳ Describe your role in the team and identify your strengths and weaknesses.	✳ You must recognise and explain your own strengths and weaknesses when working in the team. ✳ You must explain how your contribution to the team affected the success of assessment foci 6.2, 6.3 and 6.4. ✳ You must suggest how your contribution to the team could have been improved.
How to interpret information about a component which is given in engineering drawings and specifications. *Assessment focus 6.2.1*	✳ Using the engineering drawing of the component that you are going to manufacture, find four pieces of production information. ✳ Using the specification for the component, find more information which relates to the manufacturing of it.	✳ You must interpret and describe production information relating to the component you are going to manufacture. ✳ You must explain why it is important to have accurate information before planning the production of a component.

How to produce a plan and schedule for manufacturing a small batch of a product.
 	
Assessment focus 6.2.2
	✳ Decide on the size of the batch.
	✳ Produce a production plan which has information about:
	  ✳ how you intend to make the product
	  ✳ materials
	  ✳ tools and equipment
	  ✳ quality assurance
	  ✳ scheduling of events.
		✳ You should produce a detailed production plan and flow chart which have key milestones against which progress can be monitored.
		✳ You must justify the sequence of manufacturing operations.

How to set up and use tools and CNC equipment to process materials and produce finished products. *Assessment focus 6.3*	✳ Set up and use CNC tools and equipment to manufacture a quantity of the same component. ✳ Ask your teacher for guidance if you are unsure about working with a CNC machine. ✳ Obtain witness statements and take photographs to prove that you carried out the practical work effectively.	✳ You must prepare and handle correctly the materials which you are going to machine. ✳ You must maintain a clean and tidy working environment. ✳ You must evaluate the success of what you have been doing and suggest ways for improving the process in the future. ✳ You must set up and use the CNC machine without any help from your teacher apart from checking on any health and safety issues with them.
How to use quality control techniques to check that manufactured products conform to specification. *Assessment focus 6.4*	✳ Select a number of components from the batch which you have machined. ✳ Measure at least three dimensions on the component e.g. a diameter and record your figures in a table. ✳ Process these figures using a statistical method e.g. find the mean and range and establish if the product conforms to specification.	✳ You must record and present your statistical data in a way that allows you to explain why the product is within or outside tolerance. ✳ You must comment on how quality can be maintained during a manufacturing process.

Introduction

Most engineering products need to be maintained. Generally, the more complex a product the more maintenance is needed to keep it running. An aircraft, for example, requires regular maintenance to ensure that it is safe for flight. The consequences of not carrying out effective maintenance can be dire!

Engineers are frequently concerned with maintenance because breakdowns are non-productive, and repairs can often be costly, both in terms of replacement parts and lost production. Even worse, persistent failures can result in lost contracts and jobs.

This unit provides you with an introduction to maintenance procedures and associated activities. It will give you plenty of scope to plan and undertake simple maintenance tasks, carrying out inspection procedures, and making repairs and adjustments as appropriate. You will also learn about the statistical methods used by engineers to determine the average time before a product fails or needs repair.

When carrying out maintenance tasks it is important to make use of the appropriate personal protective equipment (PPE). As you progress through this unit you will be introduced to a variety of PPE and you will be shown how to use it. It is essential that you use it whenever the need arises.

Finally, you will be expected to carry out a risk assessment for a particular maintenance task. This activity will provide you with a valuable opportunity to become involved with the process of identifying potential hazards and putting in place appropriate control measures that will minimise their impact.

This unit is assessed by your tutor. On completing it you should:

1. Understand different types of maintenance for engineered products, plant or equipment, including the use of statistical trends.

2. Be able to carry out routine maintenance tasks and devise a maintenance procedure.

3. Understand the effects of poor maintenance, and the range of spares and replacement parts.

4. Be able to carry out a risk assessment in a maintenance environment.

Thinking points

Think about the following key points as you work through this unit:

1. What are the different types of maintenance and why are they necessary?

2. What statistical methods are used to determine the reliability of an engineered product?

3. What is inspection and why is it needed?

4. What is a maintenance procedure and what does it typically involve?

5. What are the consequences of improper or inadequate maintenance?

6. What determines the level of spare parts needed in order to repair an engineered product?

7. What are the main points of the Health and Safety at Work Act as applied to both an employer and an employee?

8. What is personal protective equipment (PPE) and why is it important?

9. What is a risk assessment, how is it carried out and why is it important?

Types of maintenance

Without effective maintenance, a product will often fail long before the end of its normal service life. Many companies, therefore, invest significant resources in equipment, personnel and facilities for maintenance. This unit will help you understand how and why maintenance tasks are performed, as well as the consequences of not performing them!

When things go wrong

When something goes wrong, do you ever think why, and how to prevent it going wrong in the future? Working with one or two other students, make a list of at least ten things that could go wrong with a bicycle on a long journey. Then say what you could do to reduce the chances of this happening.

Activity

Think about an engineered product that you use in everyday life, such as a radio-controlled model car or a DVD player. Write down a list of things that can go wrong with it. Then write down three or four things that you could do to stop these things happening, saying what needs to be done and when you need to do it.

Select one of the things that could go wrong and produce an A5-format maintenance card that clearly describes the task you have identified. Include sketches and drawings (where relevant).

Activity

Explain three different types of maintenance and describe an example of each. In each case, give reasons for using the type of maintenance including, where appropriate, any advantages, disadvantages or limitations.

MRO – maintenance, repair and overhaul

OEM – original equipment manufacturer

What is maintenance?

Maintenance is the work you need to do to keep something working in the way that you originally intended. Often this means keeping something in such a condition that it will continue to work. It can also mean repairing it when it goes wrong and then returning it to service.

Some engineering companies specialise in maintenance, repair and overhaul (**MRO**). Others companies specialise in the design and manufacture of a product. They are called original equipment manufacturers (**OEM**). Relatively few engineering companies specialise in both areas, because the skills and resources required are quite different. For example, an MRO company will not need to have a team of engineers involved with the design and development of new products. Equally, an OEM company won't usually need to have a team of service engineers to repair products.

Most engineered products need to be maintained. Generally, the more complex a product, the more maintenance is needed to keep it running. Maintenance can be critical where a product uses moving parts or fluids that might wear out. Electronic equipment with no moving parts often requires less maintenance. Think, for a moment, about a car. What parts are likely to wear out and need replacement? Your list should include:

* tyres
* brake linings
* windscreen wipers
* light bulbs
* oil and other fluids
* engine coolant.

The consequence of not replacing these items could be dire. For example, without sufficient engine coolant, the engine might overheat and eventually seize up! Maintenance tasks to be carried out on a regular basis might include:

* check (and if necessary replace) the engine oil and fuel filters
* inspect (and if necessary replace) the windscreen wiper blades
* check (and if necessary top up) the windscreen washer fluid
* inspect the tyres for wear (and if necessary replace)

* inspect (and if necessary replace) the brake pads
* check (and if necessary top up) the engine coolant.

Some of these tasks are performed on a routine basis to ensure that the car is safe and runs well. Others are more concerned with reducing the chances that the car will develop a fault and stop running.

Quite a few of the tasks involve **inspection**. Inspection is an important part of maintenance, as you will find later in this unit.

Different forms of maintenance

* **Planned maintenance** – Planned (or routine) maintenance is carried out at regular intervals. Suppliers of complex products will usually specify the maintenance required and the intervals at which it should be carried out. Sometimes these are specified in terms of elapsed time but, in other cases, they may be based on use. For example, the service interval for a car may be 12 months or 10,000 miles, whichever comes sooner. Planned maintenance is sometimes also referred to as servicing.

* **Preventive maintenance** – Preventive maintenance is conducted to keep equipment working and/or extend its life. Not all preventive maintenance is planned. For example, if the weather turns cold and you need to leave your car in an exposed place, you might decide to check the level of antifreeze to ensure the engine is protected against freezing up. In the normal course of events, this won't happen, but an expensive repair might be needed if the engine coolant should freeze!

* **Corrective maintenance** – Corrective maintenance involves repairing a product that has failed in some way. As it is usually carried out when a failure occurs, it can be considered to be a form of **unplanned maintenance.** An example is replacing a car exhaust system when it becomes noisy, or when an exhaust leak has been detected. Note, however, that a repair should also involve finding out how and why the product failed to ensure that it doesn't happen again.

* **Front-line maintenance** – Front-line maintenance is performed while a product, plant or equipment remains in service. This minimises interruption to operation but it can usually only be carried out where specialised resources are available. An example might be changing one of the solar panels on an orbiting space station. Why? Because it would be extremely expensive and disruptive to have to return the space station to Earth just to replace a solar panel!

* **Breakdown repair** – Breakdown repair is similar to corrective maintenance, but is a form of **planned maintenance**. In effect, we make a decision to run the product, plant or equipment until it breaks down and then repair it. We are usually less concerned about why and how the failure has occurred, because we have accepted that this will happen and will just deal with it.

Preventive maintenance includes tasks like checking the oil level in a car engine

Activity

What types of maintenance would you recommend for each of the following applications (note that, depending on the circumstances, more than one type of maintenance might be appropriate):

(a) a bench pillar drill

(b) a large passenger aircraft

(c) the turbine generators in a hydroelectric power station

(d) a set of traffic lights

(e) the engines of a deep sea fishing vessel.

Just checking

* What is maintenance and why is it necessary?
* What is a MRO?
* What is an OEM?

Statistical methods

Engineers often used calculations and statistical methods to determine production trends and estimate the likelihood that a product or equipment will not fail. These methods can be extremely powerful and make a significant contribution to improvements in product quality and overall cost-effectiveness.

Examining trends

Production engineers are often concerned with how many manufactured products or component parts are faulty, or become **early failures**, due to a manufacturing defect. These can usually be rectified by modifications or improvements to the production process or by ensuring that materials and components used are of better quality.

Detecting trends involves gathering statistical data on the number of products manufactured and the number of defective products or early failures. The number of defects detected often varies over time, because different batches of materials and components are used. For this reason hourly, daily, weekly or monthly production data is gathered; then it is examined to detect the proportion of failures or which batches exhibit unacceptably high failure rates. It is then possible to examine causes and make modifications and improvements to rectify problems.

For example, consider the production of DVD drives manufactured by a particular company in the second quarter of 2007 shown here:

Second quarter 2007	April	May	June	Second quarter totals
Total production	29,700	45.1150	51,000	125,850
Defective	495	525	451	1.471
Percentage faulty	1.66%	1.16%	0.88%	1.17%

The statistics show a trend: production volume has increased in the three month period, the proportion of faulty DVD drives has fallen by about 50%. This indicates a significant improvement in quality.

Failure rate

Most engineered products are made from a large number of parts or components. Unfortunately, a failure of any one might result in the complete failure of the product. The basic **failure rates** for some common types of electrical/electronic components are given in the table.

Different parts also have different failure rates and this must be taken into account when we calculate the overall failure rate and MTBF of a product. Not only that, but failure rates can drastically increase if products and equipment are operating under conditions of environmental stress (e.g. mechanical shock or vibration, excessive temperature variation, high

Electrical wiring and connectors fitted to this Rolls Royce Trent Airbus A380 engine need to use component parts with very low failure rates

Component type	Failure rate (average failures per million hours)
Capacitor (plastic film type)	0.002
Crimped joint	0.0003
Filament lamp	4
Fuse	0.01
Motor	1.6
Resistor (oxide film type)	0.001
Soldered joint	0.002
Switch	0.001
Transistor (bipolar junction type)	0.0002
Welded joint	0.00005

Basic failure rates for some common electrical/electronic components

Component type	Typical weighting factor
Laboratory environment (benign)	1
Commercial/domestic environment	2 to 4
Industrial environment	4 to 10
Moving vehicle (car, bus or lorry)	8 to 12
Civil aircraft	18 to 25
Helicopter	35 to 45
Military combat aircraft	25 to 55

Weighting factors for different types of environment

Activity

Look at the tables.

1. What type of electrical/electronic joint is (a) most reliable and (b) least reliable. Suggest reasons for this.

2. What type of operating environment is (a) least demanding and (b) most demanding. Suggest reasons for this.

Activity

ACE Microcontrollers has asked you to advise them on problems they have identified with the automated production of a computer board. They have provided you with the following production data:

		Week 33	Week 34	Week 35
Machine A	Total production	375	282	467
	Faulty units	22	12	10
Machine B	Total production	230	312	411
	Faulty units	4	10	27

Calculate the percentage of boards that fail for each period and for each machine. Then compare the statistics and identify any trends. What do they indicate? What can you say about the two machines? What recommendations would you make and what further investigations should the company undertake?

Activity

1. A particular type of heater has a failure rate of 3.9% per 1,000 hours. What is the heater's MTBF and, on average, how long will it be before the heater needs repair?

2. The service life (in hours) of a batch of 20 low-energy lamps is as follows: 3223, 3001, 5144, 9581, 5376, 5585, 8236, 6959, 2176, 3651, 4889, 4615, 8657, 5900, 6065, 6882, 7125, 7728, 7291, and 6337. Determine the average service life (MTTF) for this batch and use it to determine the failure rate.

levels of humidity, etc.). When determining MTBF for real applications, additional weighting factors are normally applied to take these into account, as appropriate.

Just checking

* Why are production statistics useful and what trends can they indicate?

* What does an early failure usually indicate?

* What is the relationship between failure rate and MTBF/MTTF?

Routine maintenance tasks

Adequate maintenance is essential to keep engineered products and equipment working properly. It is also essential for manufacturing and processing plants and in the generation of energy. In all of these applications, maintenance is a planned, ongoing and routine activity.

Routine inspection and adjustment

Routine maintenance is designed to ensure a product, plant or equipment operates correctly without a breakdown. It is maintenance that is planned and preventive as it reduces the chance a fault will occur later. Many products benefit from routine maintenance in the form of inspection and adjustment. Inspection may identify parts or consumables, such as grease or oil, that may need replacement. Inspection may not just be visual – it can involve smell, touch and sound!

Consider the maintenance procedure recommended for a portable mains-operated hammer drill.

1 Before every use

∗ Check that mains lead and plug are not damaged.

∗ Check for any damaged parts including chuck, handle and controls.

2 After every use

∗ Wipe hammer drill using a clean dry cloth.

3 After every *five hours* of use

∗ Remove and inspect chuck. Remove any chips or dust that may be lodged and check that the jaws of the chuck move and engage properly.

∗ Open grease box using a pin spanner and check grease level. Top up using a general purpose lithium-based grease.

5 After every *20 hours* of use

∗ Inspect mains plug and lead for damage (particularly at cable grip).

∗ Check internal plug wiring and that the correct fuse has been fitted.

∗ Tighten any screws that may have become loose.

∗ Remove brush cover using a cross point screwdriver. Unclip and withdraw each brush. Inspect and clean each brush.

∗ Replace if damaged or worn. Clean commutator segments with a soft brush and isopropyl alcohol solution

∗ Replace brushes and brush cover, check hammer drill operates correctly.

Maintenance procedures

Maintenance procedures are documents that detail the tasks to be performed as part of planned maintenance. They usually comprise:

∗ sequence to follow (in a logical order)

∗ timescale (or intervals) at which the tasks should be performed

∗ tools, materials, test/measuring equipment, etc. that will be required

∗ documentation required (service manuals, exploded diagrams, etc.)

∗ safety practices (handling precautions, safety equipment, and disposal procedures)

∗ diagnostic routines (such as test procedures, calibration procedures, etc.).

Keeping it running

Have you checked the tyres or changed a wheel on a bike or car? Have you replaced a lamp or fuse, changed a plug or replaced a battery? These are all essential maintenance tasks that you perform on a regular basis. With one or two other students, make a list of at least five tasks you would perform to ensure a bicycle is fit for use. Illustrate your answer with a flipchart diagram showing the parts that need checking and adjustment.

Activity

Devise a comprehensive maintenance procedure for the tyres fitted to a car. Ensure that your procedure is presented in the form of a logical sequence of tasks and that it gives full details of the timescale, tools and equipment required, safety precautions, inspection and testing requirements, references to other documents (such as the service manual for the car), as well as the visual checks and inspection procedures that should be carried out. Hint: You may find it useful to carry out some preparatory research.

Inspection – examination by sight, touch, smell and sound (as appropriate) to determine the functional state of a product, process plant or equipment

Maintenance procedure – the logical sequence of tasks that should be performed when carrying out routine maintenance

Inspecting the brushes and commutator for wear or damage

Inspecting the plug (checking the cable grip, internal wiring and that the fuse is correctly rated)

The maintenance procedure for the cooling fan unit fitted to the central processing unit (CPU) on a computer might be as follows.

1 After 1,000 hours of operation (or 12 months, whichever comes first) run PassMark burn-in software and select temperature checking routine.

2 If CPU temperature is abnormally high (see p. 5 of manual) CPU cooling unit and integral fan (part number 12-189-1032) may need replacement.

3 Switch 'off', disconnect from the supply and access system board by removing external case.

4 Gain access to motherboard area around CPU by removing adapter cards or cables that may restrict access (see Fig. 3 in manual).

5 Ensure that you observe the safety and static precautions at all times.

6 Locate CPU and ensure there is sufficient room to work all around it (you may have to move ribbon cables or adapter cards to gain sufficient clearance for extraction tool).

7 Disconnect fan cable from motherboard header (SK-152) (see Fig. 7 in manual).

8 Remove CPU fan cooling assembly by unclipping and withdrawing it using correct cooling unit extraction tool (part number 77-500-1001).

9 Check replacement CPU cooling assembly is correctly located over CPU before pressing down firmly and locking into place.

10 Connect fan cable to motherboard header (SK-152) (see Fig. 7 in manual for location).

11 Re-assemble system (replacing any adapter cards and cables removed to gain access or clearance around CPU).

12 Reconnect system, switch on and test.

13 After 15 minutes, check CPU temperature is within specification by entering system folder and running CPU temperature checking routine in PassMark burn-in software.

14 Repeat temperature check after a further 90 minutes of operation.

Effects of poor maintenance

Understanding the implications of poor maintenance is important because improper maintenance can have serious consequences, including the very real risk of personal injury. Maintenance procedures should therefore only be carried out by those who are suitably qualified and experienced, using appropriate tools and resources.

Impact of improper maintenance

When effective maintenance procedures are not in place, the consequences might include one or more of the following.

* **Dissatisfied and disaffected customers** – When a product fails, it can reflect on the original equipment manufacturer (OEM) as well as the maintainer of the product.

* **High cost of making repairs** – For example, most vehicle manufacturers recommend that the engine cam belt should be replaced after an interval of about 40,000 miles. If this procedure is not carried out and the belt subsequently fails, it might be necessary to replace the entire engine. This can be extremely expensive – the cost can be even greater than the value of the car!

* **Safety hazards, injury and risk to life** – For example, when a poorly maintained electrical appliance fails, there can be a risk of fire or electric shock.

What might happen if it goes wrong?

When things go wrong, the consequences can sometimes be serious. Working with one or two other students, think about what might happen if a car is not maintained correctly. List three maintenance tasks which, if not properly performed (or not performed at all) could potentially result in death or injury.

Case Study: G-YMME

On the 10th June 2004, a Boeing 777 aircraft, registration G-YMME, took off from London Heathrow airport bound for Harare in Zimbabwe. Here is an account of its flight and subsequent return to Heathrow.

As the aircraft accelerated down the runway and started to climb, the crew of another aircraft, waiting at the runway holding point, reported that they could see a trail of smoke from the rear of the departing aircraft. They also reported a smell of fuel vapour. This was confirmed by further reports from the ground that indicated that there was a two-mile trail of fuel vapour from the rear of the aircraft. This information was communicated by radio to the Captain and First Officer of G-YMME, who reported that they could not see any abnormal indications on the flight deck, nor could they see anything from inside the aircraft.

From the information that they had received, the crew of G-YMME decided to dump fuel to reduce the weight of the aircraft and then return to Heathrow for an emergency landing. To keep the brake units as cool as possible, and thus avoid risk of igniting any leaking fuel, the landing was made with minimum braking; after which it was taxied back to a stand where the passengers disembarked.

An engineer made a preliminary inspection of the aircraft at the stand. He noticed that there were a few drips of fuel on the left main landing gear but none on the ground. After opening the left main gear door, he detected a distinct smell of fuel. An inspection inside the gear bay door revealed that the centre fuel tank purge door was not in place. Instead, it was hanging on a lanyard, together with a plastic bag containing the screws that normally held the purge door in place. During inspection and maintenance, the fuel tank purge door is opened to avoid the accumulation of fumes in the tank; but during flight and at all other times, the door should have been closed.

From what you've just read, what do you think was the cause of the emergency? Clearly someone had not replaced an important component when the aircraft was being maintained. But why was the tank purge door not replaced and how did this happen? And, if the purge door has been missing for some time (as was probably the case), why had fuel not leaked out before?

Cases like this are always investigated by the Air Accidents Investigation Branch (AAIB) as well as by the aircraft operator (in this case, British Airways). The AAIB carried out a thorough investigation which made a number of conclusions and recommendations. Here is what they found.

* The fuel tank leak was caused by fuel escaping through an open fuel tank purge door inside the left main landing gear bay and the tank had been closed without ensuring that the purge door was in place.

Case Study: G-YMME continued

* When the purge door had been removed (some time previously), a defect job card should have been raised for the removal and refitting of the door, but this had not been done.

* The purge door was not mentioned in the Aircraft Maintenance Manual (AMM) procedures for purging and leak-checking the fuel tank (so no record was made of its removal) and, with no record of the purge door removal, the engineering checks for visual leaks did not include the purge door.

* The fuel quantity required to check the tank purge door for leaks was incorrectly stated in the AMM as 32,000 kg instead of the 52,163 kg required (which meant that the fuel level was still below the open purge door – the leak would probably have been detected if more fuel been used during the leak check procedure)

* A significant quantity of fuel had leaked from G-YMME's tank because the large fuel load that it was carrying for the flight to Harare moved towards the rear of the fuel tank during the take-off run and initial climb. The leak was not in evidence on landing after the aircraft had dumped most of its fuel load.

* There was a low level of awareness of the existence of the tank purge door amongst the engineering maintenance staff, due in part to the absence of references to the procedure in the AMM.

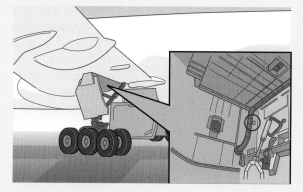

Location of the fuel tank purge door high up at the rear of the centre fuel tank and accessed from within the landing gear bay

Activity

Having read the case study of G-YMME, list three recommendations that you would make concerning the fuel tank maintenance procedures for a Boeing 777 aircraft. Explain why you have made each recommendation and say who should be responsible for carrying them out.

Activity

A car hire company has asked you to advise them on tyre maintenance. Describe and explain two likely implications of not having an effective tyre maintenance procedure. In the context of this company, justify the use of the tyre maintenance procedure that you devised earlier (page 171) and explain the benefits of proper tyre maintenance. Present your work as a brief written report.

Boeing 777 G-YMME in flight

Just checking

* Why is proper maintenance important?
* What are the implications of poor maintenance?

To repair a product that has gone wrong, you usually require one or more spare parts. These need to be available when you carry out a maintenance procedure. Failure to have these available can be expensive and inconvenient and is not a good idea if you are aiming to satisfy your customers!

The need for spare parts

Some parts and components can be prone to sudden failure. For example, light bulbs can suddenly fail and need replacement. Other parts and components can gradually wear out. For example, vehicle brake linings slowly wear out until they no longer work effectively and must be changed. Regardless of whether they suddenly fail or gradually wear out, both types of component will eventually need replacement. The important question is how many replacement parts or components will be needed:

* too few and you might not be able to carry out a repair when your stock of spare parts becomes exhausted;

* too many and you will have cash locked up in parts or components that might not be used for a long time.

There could also be cost and inconvenience in providing storage for parts that might not be needed for some time.

Calculating stock levels

What's needed is some means of predicting the minimum number of replacement parts that you need to have available, so that you avoid the risk of them running out when you need them. You can easily calculate the **maximum stock level** (MSL) that you need to hold. To understand how this works, suppose that you need to fit a particular replacement part every day and that the lead time from your supplier is five days (lead time is the time that it takes from ordering the component to it arriving on the shelf ready for you to use). This might suggest that you will need to hold a minimum of five parts in stock and, when your stock level falls to just five parts, you need to reorder. Putting this into a formula we arrive at:

MSL = HDU × ALT

where MSL = maximum stock level, HDU = historical daily usage, and ALT = average lead time.

In this example, because we use one part every day, the HDU is 1 and ALT is 5. Therefore:

MSL = 1 × 5 = 5 parts.

When the number of parts held in stock has fallen to the minimum stock level an order for replacement parts should be generated. The reorder quantity should be at least equal to the maximum stock level.

Worst case conditions

At first sight, this strategy for calculating the maximum stock level might look OK. However, suppose that on day five and day six you need to have two replacement parts, not just one. You would run out of stock two days early. And what if your supplier runs out of stock or can't manufacture the parts quickly enough, and the lead time increases from five to seven days? This would make the problem even worse.

You need to take into account the worst case scenario for both HDU and ALT. In this event you would need some additional **safety stock**. You can calculate this from:

SSL = MHDU × (MHLT − ALT)

where SSL = safety stock level, MHDU = maximum historical daily usage, and MHLT = maximum historical lead time.

So, taking into account the most difficult set of circumstances, where you know that the MHDU is 2 and the MHLT is 7, you arrive at:

SSL = 2 × (7 − 5) = 4 parts.

The **reorder stock level** (RSL) is the sum of the MSL and SSL:

RSL = MSL + SSL = 5 + 4 = 9 parts.

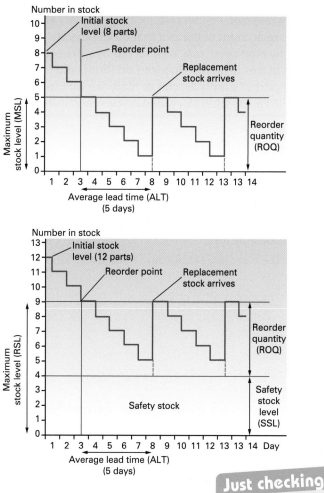

Number in stock

Maximum stock level (MSL)

> **Stock level** – the number of spare parts held in stock and immediately available
>
> **Maximum stock level** – the number of parts that need to be held in stock given the rate of use and the supplier's lead time. Replacement parts must be ordered in quantities equal to the maximum stock level
>
> **Lead time** – the time it takes from ordering the component to it arriving
>
> **Safety stock level** – the additional number of parts to be held in stock to cope with worst case conditions
>
> **Reorder stock level** – the number of parts held in stock that trigger the need to reorder a quantity of parts from the supplier equal to the maximum stock level

Activity

An air force forward maintenance unit holds a stock of aircraft batteries. The normal lead time for a replacement is 14 days but this can go up to 21 days under combat conditions. On average, one aircraft battery is replaced every seven days. However, under combat conditions one battery might need to be replaced every other day. What is the safety stock level for aircraft batteries? At what stock level should the unit re-order?

Just checking

* What is the maximum stock level and how is determined?
* What are lead times and why are they important?
* What are worst case conditions and why are they important?
* How is the reorder quantity calculated?

Keeping safe

Thought shower – How aware are you of what the law requires in terms of Health and Safety? Do you know how the law affects your work in school or college? Does your school or college have a Health and Safety Policy and who is responsible for it? What happens if there's an accident? Who needs to know about it and how should it be reported? Working in a group with two or three other students, discuss the answers to these questions.

Personal learning and thinking skills

Visit the Health and Safety Executive (HSE) website at www.hse.gov.uk. Locate the section on transport and go to the page on vehicle selection and maintenance. Make a list of ten HSE recommended features for vehicles that will be used in the work place. Present your work as an A3 poster that can be prominently displayed in a vehicle bay.

Hazard – A condition that constitutes a risk to health and safety.

Manual operation – An operation that is performed by hand, such as lifting, stacking, loading and unloading.

The Health and Safety at Work Act (1974) makes both the employer and the employee equally responsible for safety; both are equally liable to be prosecuted for violations of safety regulations and procedures. It is the legal responsibility of every employee to take reasonable care of his or her own health and safety. The law expects you to act in a responsible manner so as not to endanger yourself nor be a danger to others. This is particularly important for anyone working in engineering, where hazards exist that could potentially harm you as well as others.

Health and Safety at Work Act

All work activities are covered by the Health and Safety at Work Act 1974. The Act encourages individuals to be aware of their own safety and take responsibility for their own actions. You need to be aware that the Health and Safety at Work Act applies to people, not to premises. The Act covers all employees in all employment situations. The precise nature of the work is not relevant, neither is its location. The Act also requires employers to take account of the fact that other persons, not just those that are directly employed, may be affected by work activities. In addition, the Act also places certain obligations on those who manufacture, design, import or supply articles or materials for use at work to ensure that these can be used safely and do not constitute a risk to health.

The duty of the employer

Under the Act, it is the duty of the employer to ensure, so far as is reasonably practicable, the health, safety and welfare of all employees. The employer also needs to ensure that all plant and systems are maintained in a manner so that they are safe and without risk to health. The employer is also responsible for:

* the absence of risks in the handling, storage and transport of articles and substances
* instruction, training and supervision to ensure health and safety at work
* maintenance of the workplace and its environment to be safe and without risk to health
* providing, where appropriate, a statement of general policy with respect to health and safety and to provide arrangements for safety representatives and safety committees
* conducting his or her undertakings in such a way as to ensure that members of the public (i.e. those not in his or her employment) are not effected or exposed to risks to their health or safety
* giving information, to persons who are not employees, about the way in which undertakings are conducted that might affect their health and safety
* the safety of members of the general public (including clients, customers, visitors and passers-by).

The duty of the employee

Under the Act, it is the duty of every employee, whilst at work, to take all reasonable care for the health and safety of themselves and other persons who may be affected by what they do, or fail to do. They are required to:

* co-operate with the employer to enable the duties placed on the employer to be performed

* have regard of any duty or requirement imposed upon their employer, or any other person, under any of the statutory provisions

* not interfere with, nor misuse, anything provided in the interests of health, safety or welfare under any of the relevant statutory provisions.

Safety hazards

You will already be well aware from Unit 4 Topic 2 that many engineering processes (e.g. welding, casting, forging and grinding) can be potentially dangerous. In addition, some involve hazardous materials and chemicals. You might not know that many processes and maintenance tasks, usually thought of as 'safe', can become dangerous when carried out using improper tools, without correct preparation or without observing basic safety procedures.

Maintenance tasks that are particularly hazardous include:

working at height	welding, brazing, soldering and grinding
working in an enclosed space	working with compressed air
working underground	working with paint and other coatings
working on electrical equipment	working with machines and power tools
working with gas or chemical supplies	working with rotating machinery.

Hazards are associated with the day-to-day operation and maintenance of many engineering systems. These include those that use electrical supplies (a.c. and d.c.), compressed air, fluids, gas, or petrochemical fuels. They also include systems that are not, themselves, particularly hazardous but which are used in hazardous environments (for example in mining or in the oil and gas industries).

Operations such as refuelling an aircraft or motor vehicle can be dangerous in the presence of naked flames or if electrical equipment is used nearby. A static discharge, for example, can be sufficient, in the presence of petroleum vapour, to cause an explosion. Similar considerations apply to the use of mobile phones or other transmitting apparatus in the vicinity of flammable liquids.

Working at height in this aircraft hangar requires particular care

A well organised tools store

Warning signs and personal protective equipment

Warning signs are widely used in engineering maintenance to alert personnel and others to the presence of a specific hazard or dangerous condition. They are used to indicate fire exits, machine stop switches and other 'safe conditions'. Also, you need to be aware of the use of personal protective equipment (PPE) such as overalls, safety boots or shoes, eye and ear protection. The use of PPE should become second nature in your career as an engineer.

Warning signs

The European Safety Signs Directive 92/58/EEC is designed to standardise safety signs across the whole of the European Community. It is implemented in the UK by the Health and Safety Executive (HSE) in the Health & Safety (Safety Signs & Signals) Regulations 1996. The European Safety Signs Directive requires that:

✳ employers use safety signs in situations where there is a risk that cannot be controlled by other means

✳ safety signs should make use of pictures (or **pictograms**) to convey meaning (text should only be used to convey extra information).

Appropriate warning and prohibition signs should be prominently displayed in the workplace and also when carrying out maintenance tasks on clients' or customers' premises.

There are five main types of warning sign.

✳ **Prohibition signs** – things that you must not do, e.g. No Smoking.

✳ **Warning signs** – signs that warn you about something that is dangerous, e.g. 'Danger: High Voltage'.

✳ **Mandatory signs** – signs that indicate things that you must do, e.g. 'Eye protection must be used'.

✳ **Safe condition signs** – signs that give you information about the safest way to go, e.g. 'Fire Exit'.

✳ **Fire signs** – signs that indicate the location of fire fighting equipment, e.g. 'Fire Point'.

Note that different colours are used to make it easy to distinguish the types of sign. Safe condition signs use white text on a green background, mandatory signs use white text on a blue background, and so on. It is essential that you familiarise yourself with the different types of sign and what they mean!

Prohibition signs
White text on a red background

Warning signs
Black text on a yellow background

Mandatory signs
White text on a blue background

Safe condition signs
White text on a green background

Fire signs
White text on a red background

Five different types of sign

Flammable material

Oxidising agent

Caustic material

Electric shock hazard

Electric shock hazard

Compressed gas

Fork lift truck

Toxic material

Unsafe roof

Radioactive material

Explosive hazard

Explosive hazard

A selection of warning signs

Typical use of warning and mandatory signs on a power guillotine

Personal protective equipment

Personal protective equipment (PPE) must be used when performing many engineering maintenance tasks. It includes clothing and footwear, as well as special equipment for eye protection, ear protection, etc.

Clothing and footwear

Suitable and unsuitable working clothing for use in an engineering workshop is shown in the picture. Overalls or protective coats should be neatly buttoned and sleeves should be tightly rolled. Safety shoes and boots should be worn (not trainers!). Overalls and protective clothing should be sufficiently loose to allow easy body movement but not so loose that they interfere with engineering tasks and activities.

Special equipment

Some processes and working conditions demand even greater personal protection. In such cases PPE can include safety helmets, ear protection, respirators and eye protection worn singly or in combination. Such protective clothing must be provided by the employer when a process demands its use. Employees must, by law, make use of such equipment

Short hair — Long hair

Sleeve tightly rolled — Sharp tools

Button missing

Buttons fastened — Loose cuffs

Hole in pocket

Overalls correct length — Overalls too long

Safety boots — Lightweight shoes

Right and safe — **Wrong and dangerous**

Pictogram – a picture that conveys meaning and can be widely understood. Pictograms are often used as an alternative to text, particularly where many different languages may be used

Personal protective equipment (PPE) – clothing, footwear and other items that can be worn or carried to avoid risks and minimise hazards

Unit 7 Assessment Guide

In this unit, you will start by investigating some of the maintenance procedures used by industry to keep machinery and equipment in good condition and operating to specification. These need to be carefully planned and documented so that when engineers service machinery it can be done at times which will have the least impact on production: for example, when machine operators are on holiday or between shifts.

Time Management

Manage your time well as this unit has a number of different components that will have to be researched. Ensure that you keep your work safe and that any work in electronic format has a secure and safe backup.

Be well organised. This is your chance to show that you are an independent enquirer, creative thinker, reflective learner, self manager and effective participator and therefore will contribute towards achievement of your Personal learning and thinking skills.

Plan ahead for your work experience and make a checklist of things that you need to investigate so that you make the most efficient use of your time.

Useful Links

Make good use of your work experience to find out as much as possible about issues that are relevant to your coursework. You should arrange to work and meet with people in the maintenance department of the company.

Things you might need

Your work should be in the form of an A4 process portfolio and should be presented in an e-format. Your teacher should give you access to the required software to enable the correct presentation.

The data needed for assessment focus 7.1.2 could be found when you are on work experience.

Remember you will have to carry out practical activities in order to complete assessment focus 7.2.1. Make sure you obtain witness statements from your supervisor and photographs showing you working correctly and safely.

A digital camera or mobile phone would be useful, so that you can record evidence of practical activities.

You would benefit from access to a factory maintenance technician who works on plant and equipment.

Remember to maintain a focus on the fact that equipment breakdowns are unacceptable to industry: lost production costs money, and products which are not delivered on time may alienate the customer.

How you will be assessed		
What you must show that you know	**Guidance**	**To gain higher marks**
That there are different types of maintenance technique used in industry. *Assessment focus 7.1.1*	✳ Describe two different types of maintenance procedure used in an engineering factory.	✳ You must be able justify why why a particular maintenance procedure is being used.
That statistical methods are used to predict when a piece of equipment is going to break down or go out of tolerance. *Assessment focus 7.1.2*	✳ Find the maintenance log for a piece of equipment (e.g. a machining centre) and put together some statistics e.g. drive motor brush change interval. ✳ Investigate why statistical maintenance data for a given piece of equipment is important.	✳ You must investigate the failure rate of a piece of equipment and be able to calculate the mean-time-to failure (MTTF). ✳ You must explain how statistical data can be used as an aid to determining equipment reliability.
How to carry out routine maintenance on engineering equipment. *Assessment focus 7.2.1*	✳ Carry out maintenance which involves routine adjustment and replacement of a consumable part. ✳ Obtain witness statements and take photographs to prove that you carried out the practical work effectively.	✳ You must read the maintenance procedure for a piece of equipment and: assess where maintenance is needed; carry out routine adjustments and servicing; replace a consumable item. ✳ You must must prove that you can identify some feature of a machine which requires maintenance.
That you know how to create a maintenance procedure for a piece of equipment. *Assessment focus 7.2.2*	✳ Write a maintenance procedure for a piece of equipment which covers: tools required; sequence of operations; safety practices; documentation.	✳ You must write a detailed maintenance procedure for a different piece of equipment and which includes: tools and test equipment; sequence of operations; diagnostic routines; safety practices documentation/timescale.
That if equipment is not properly maintained it will malfunction and may develop faults which cause it to break down. *Assessment focus 7.3.1*	✳ Find two pieces of equipment which require regular maintenance and explain what will happen if this is not carried out correctly or at the proper time interval.	✳ You must find two examples of equipment failure caused by poor maintenance. Explain the consequences and how the problem could have been avoided.
Why it is important for a maintenance department to carry a sufficient stock of spares and replacement parts. *Assessment focus 7.3.2*	✳ Find the maintenance schedule for a piece of equipment and make a list of the spare/replacement parts which need to be kept in stock.	✳ You must describe what happens if the stock of replacement parts is not kept at the correct level. ✳ You must explain how statistical data is used to control stock levels.
Why maintenance operations need to be risk assessed. *Assessment focus 7.4*	✳ Using standard documentation risk-assess a maintenance task.	✳ You must carry out a detailed risk assessment which includes information about: personal protective equipment; safety regulations; warning signs.

Introduction

Innovation is the process of developing and implementing a completely new product or service. Innovation involves putting new ideas into practice, turning them into reality through creativity, as well as research, investigation and development that is frequently based on the application and synthesis of existing knowledge.

An ability to innovate is important for anyone who wishes to develop new products and expand into new markets. This unit will introduce you to the skills and techniques used by innovators. The unit introduces the various means by which money can be raised in order to finance the development, manufacture and marketing of new products.

You will learn how to protect your work by asserting your design rights and through a formal patent application. This is important as it can maximise your ability to fully exploit your ideas and further develop them into successful commercial products. You will learn how successful entrepreneurs have responded to this challenge and, through case studies, how they developed their ideas into products that we all recognise and use.

The unit will also introduce you to a variety of engineering materials and processes, as well as the impact that they might have on the environment and society.

This unit is externally assessed by means of a written test paper. On completing the unit you should:

1. Know about the intellectual property within engineering.

2. Understand the role of research, development and raising finance when designing engineering products.

3. Know about developments in materials and processes on products.

4. Know about the effects of engineering technologies in the home, workplace or built environment.

5. Know about the environmental and social impact of engineering, and sustainability of resources.

Thinking points

Think about the following key points as you work through this unit:

1. What is innovation and why is it important?

2. What is intellectual property and how can a design be protected?

3. What is a trade mark and what is copyright, and to what do they apply?

4. What is research and development, and why is it needed?

5. What sources of finance are available for the development of new engineered products and services, and how do they work?

6. What are the main classes of materials, what are their properties and what are they used for?

7. What processes are available for use with engineering materials and what do they involve?

8. What is new technology and why is it important in engineering?

9. What are the effects of engineering activities on society and the environment, and how can adverse effects be minimised?

Engineering innovation

The ability to come up with new and revolutionary ideas for new products can lead to spectacular success in a global market. Such innovation needs a fertile and creative mind coupled with a determination to bring a new product to market. It's important that would-be entrepreneurs and inventors turn their new ideas into reality.

Got an idea?

Many successful inventions start from trying to satisfy a need that is not fulfilled. Working with two or three other students, agree on one labour-saving device for which no existing solution is available. Then think of three things that you would need to do to take your idea further.

Knowledge-based organisation – an organisation that uses a fund of knowledge for its day-to-day operation.

Implicit knowledge – knowledge we hold in our minds but is not written down

Explicit knowledge – knowledge that exists on paper or in electronic form

What is innovation?

Innovation is about bringing something new into the world, which is better than what currently exists. It most often refers to a new product, but can also apply to services, manufacturing and management processes, or the design of an organisation. It can also include improvements to the efficiency or effectiveness of existing products, processes or services.

Innovation involves creativity. It involves taking new ideas and turning them into reality through invention, research and the development of new products and services. Successful innovation requires:

* an ability to think creatively
* a wide knowledge of existing solutions and technologies
* an ability to think unconventionally ('outside the box')
* an ability to seek, apply and experiment with new possibilities and techniques
* an ability to communicate ideas and to enthuse other people
* an ability to protect and promote ideas within the legal framework.

Knowledge and innovation

Knowledge and innovation are often linked. Knowledge is usually considered to be something built up in the mind of individual people, resulting from their interaction with the world around them. However, organisations also have knowledge – within the minds of employees but also in the form of paper and electronic records. A good example of a **knowledge-based organisation** is a school, college or university. However, many large organisations and corporations now regard themselves as stores of knowledge and are beginning to realise that this fund of knowledge can be a huge asset.

Knowledge is often considered to be either implicit or explicit. **Implicit knowledge** is held in a person's mind and doesn't need to be turned into words. For example, we know that a kettle gets hot and gives off steam, so we observe caution when we approach one. **Explicit knowledge** is knowledge that has been written down, drawn, or otherwise expressed to communicate it to other people.

The Dyson vacuum cleaner

Case Study: The Dyson vacuum cleaner

Sir James Dyson is a well-known industrial designer. He is best known as the inventor of the Dyson vacuum cleaner and is currently said to be worth more than £1 billion. More than 30 years ago, James became frustrated with the performance of his conventional vacuum cleaner. He found that the main problem was that the dust picked up was clogging the bag, so the cleaner lost suction and became less efficient. James's idea was to improve the performance of the vacuum cleaner by using the principle of 'cyclonic separation'. This was the principle used in the air filter of the spray booth in the factory where he was producing the Ballbarrow (an innovative wheelbarrow that uses a ball rather than a wheel).

After five years of development in which James produced several thousand prototypes, the design for the Dyson 'G-Force' bag-less vacuum cleaner was complete. Unfortunately, no manufacturer or related distributor was willing to launch his product in the UK, because it was felt that the new invention would damage sales of existing vacuum cleaners and their replacement bags. James then turned to the overseas market, initially launching the 'G-Force' bag-less vacuum cleaner in Japan. To protect his invention, James obtained his first United Sates patent in 1986 (U.S. Patent 4,593,429).

In 1993, frustrated by the ongoing resistance to his idea from manufacturers, James decided to set up his own vacuum cleaner factory and research facility in Wiltshire. Dyson engineers discovered that a smaller diameter cyclone gave greater centrifugal force. This led to a way of getting 45% more suction than the original dual cyclone and removing more dust, by dividing the air into eight smaller cyclones and incorporating this principle into later bag-less cleaners. Since then, the Dyson product range has become firmly established and now outsells many of the products from companies that rejected his original idea.

What an innovator needs to consider

* What is the need that I am trying to satisfy?
* What are the existing ways of satisfying this need and why are they unsatisfactory?
* What knowledge do I need and where can I find it?
* What technologies could be exploited and how fully developed are they?
* What other ways are there of solving this problem?
* Where is the market for the solution and how can I reach it?
* How can I promote my idea and raise the capital I need for developing it?
* How can I protect my idea so that it does not fall into the hands of others?

Later in this unit, you will explore these in much greater detail.

Activity

Find out about the Dyson Ballbarrow. What is the advantage of this compared with a conventional wheelbarrow? In what later products is the principle of the Ballbarrow employed?

Activity

Find out about the inventor of the clockwork radio, Trevor Baylis.

1 What need was Trevor trying to satisfy?

2 What did he do to satisfy that need and how well did he do it?

3 What did he use to test the principle behind his invention?

4 What event proved to be instrumental in bringing Trevor's invention into the public eye?

5 Which company first manufactured his invention?

6 What other organisations were involved in the design and manufacture of Trevor's invention?

Personal learning and thinking skills

Trevor Baylis once said 'The key to success is to risk thinking unconventional thoughts. Convention is the enemy of progress'. How did Trevor apply this concept to his invention? Discuss this with your group and suggest at least two other ways in which the original need could have been satisfied.

Just checking

* What is innovation and why is it important?
* What is knowledge and where is it found?
* What things does an innovator need to consider?

UNIT 8 EXPLORING INNOVATION, ENTERPRISE AND TECHNOLOGICAL ADVANCEMENTS **185**

Designs and patents

Being able to protect a design or invention provides you with an advantage over your competitors and means that you can publicise your design without the risk that anyone else might steal your ideas. You might also want to sell or license your ideas to others, so that you can derive financial benefit.

Intellectual property

If you have an idea that is new it becomes your **intellectual property** (IP) and is protected by UK law. IP applies to a great many things, including ideas that are written down as well as sketches, drawings, diagrams... anything that has been invented by you.

Countries have laws to protect IP for three main reasons:

✳ to recognise the moral and financial rights of creators in their creations, as well as the rights of the public who might want to access those creations

✳ to promote, as a deliberate act of government policy, creativity and the spreading of ideas and inventions

✳ to encourage economic and social development through fair trading.

Intellectual property is traditionally divided into two main types:

✳ industrial property, including inventions (patents), trademarks, and designs

✳ copyright, which includes writing and artistic works.

Designs

Design relates to the physical appearance of a product, not the function or operation of the product. Features that correspond to the design and physical appearance of a product include size, shape, texture, colour, lines and contours and materials.

Several forms of IP could apply to a design.

This tape dispenser is a registered design

A **UK design right** is an **automatic right** (it does not have to be registered) that prevents other people from copying your design. It is not a complete right as it covers only the three-dimensional aspects of the item and does not protect the surface decoration of the product or any two-dimensional pattern, such as a wallpaper or carpet design. UK design right lasts for up to 15 years from creating the design.

A **UK Registered Design** is protected throughout the UK. The protection lasts initially for five years and can be renewed every five years for up to 25 years. To register the design, you need to apply to the UK Design Registry and pay a fee. After that, you are allowed up to one year from first disclosing your design publicly before you have to register it. A similar protection exists within EU member states – **Registered Community Design** (RCD), which also offers protection for up to 25 years.

Unregistered Community Design offers protection from copying the design on any item. Protection lasts for three years after the design has been made available to the public and covers all EU countries. Like a UK design right, it is an automatic right which you don't need to apply for.

Intellectual property (IP) – property created by intellectual or creative activity, including patents, trademarks, copyright and designs. Like all property, if you own it you can rent it, lease it, license it, give it away or sell it

Automatic right – IP right not requiring a formal application or payment of a fee

Patent – exclusive right to use an invention commercially in return for disclosing it and payment of fees

Patent application – the documents you need to file, giving details of your invention

You can register a design if you can show that it is:

* new (not the same as any design already available to the public)
* individual in character (the overall impression the design gives must be distinct from any previous designs).

You cannot register a design if:

* it is more than 12 months since the design was first publicly disclosed
* the design is about how the product works, not what it looks like or how it appears
* the design includes complicated component parts that remain unseen in normal use
* it is offensive
* it involves certain national emblems and protected flags.

As with other intellectual property rights, owning a registered design means you can sell, or license someone else to use it. You have one year from first showing your design to the public to assess whether it is marketable. In this period, your design is protected by design right. If you then want to apply for registered design status, you must do it no later than 12 months from first showing the design.

Patents

If you develop a process or a product that is new or inventive and could be made, you can apply for a patent. This can protect your invention by making it unlawful for anyone, apart from you or someone with your permission, to produce, use, import or sell it. It gives you an exclusive right in any country where the patent has been granted, as long as you pay the renewal fees every year.

A granted patent becomes property and, like any other property, you can buy, sell or license it out. However, you must not reveal your invention publicly before you apply for your patent or it may jeopardise your chances of it being granted!

	Registered design right	Patent
What does it protect?	The form, shape, look and general appearance of a product	Things that you have invented
What form does the protection take?	Prevents your product from being manufactured, sold or imported by someone else	Prevents your idea from being used, manufactured or sold by someone else
Where does it operate?	Throughout the UK	Throughout the UK
How long does it last?	Up to 25 years	Up to 25 years (subject to annual renewal)

Just checking

* What is intellectual property?
* What is a design and how is it protected?
* What is a patent and how do you go about applying for one?

This Pocket PC is protected by a wide variety of registered trademarks.

Trademarks and copyright

Design rights and patents are two ways that you can protect your intellectual property. You also need to know how trademarks and copyright are used.

Trademarks

A trademark is a sign that distinguishes your products and services from others. A trademark can include words, symbols and pictures, or a combination of them. An easily recognisable trademark can also be a very effective marketing tool and many companies spend a lot of time and effort to get their trademark logos right.

Registering a trademark gives you the right to use it on products and services in certain classes. Registering the mark gives you the right to take legal action against anyone who uses your mark, or a similar mark, on products and services similar to those you set out in the registration. To be able to register a trademark, it *must*:

✱ be distinctive when used with the products and services you are applying to register it for (invented words are usually considered distinctive)

✱ be based on words, symbols and logos that would be perceived by members of the public as representing a trademark.

To be able to register a trademark, it *must not*:

✱ be the same as (or similar to) any earlier marks on the register for the same (or similar) products or services

✱ use terms that are commonly used to refer to the product or service, or to its characteristics or properties, (for example, the trademark 'Orange' would not be allowed for use with the supply of fruit and vegetables but is perfectly acceptable for a mobile phone operator!)

✱ use offensive, illegal or deceptive terms or symbols

✱ incorporate advertising slogans or other promotional material

✱ use protected emblems (for example, flags, Red Cross or Olympic symbols).

A trademark application may also be objected to by the owner of an earlier mark that is considered confusingly similar. To avoid this, existing registrations are searched. If any similar marks are found, the owners of the earlier registration are informed so that they can object.

If you use an unregistered trademark, you still have certain rights under common law and you can use the ™ symbol. However, it is easier to enforce your rights if you register your mark and use the ® symbol. The registration must be renewed every ten years. Note that it is an offence to use the ® symbol if the mark is not registered somewhere in the world. Like other IP (intellectual property) rights, a trademark can make you money if you sell, lease, or license it for use by someone else. It can be a valuable asset.

Copyright and trademark information displayed in the Pocket PC's User Guide

Copyright

Copyright is used to protect books, plays, films, sound and video recordings, broadcasts, photographs, diagrams and other illustrations, software, websites, and logos. Copyright applies as soon as a work is written down, or recorded or stored in a computer memory, and it is an automatic right that you do not need to formally apply for.

To show that something is copyright, you can use the © mark followed by your name and the date. Where the copyright might be disputed, you can deposit a dated copy of your work with a solicitor or bank. If someone infringes your copyright, you can prove that you were the original creator.

Copyright is an IP right that relates to the expression of an idea, and not the idea itself. For example, anyone can make a film about a school for wizards but they must not copy parts of an existing film, or base it on a story written by someone else without their express permission.

If you use someone else's copyright material, you must get their written permission first. An exception is where an organisation represents a group of copyright owners who offer blanket licences to users, in return for royalties.

Copyright lasts for a different length of time according to the type of material. For example, the right to the layout of this book (how it looks on the page) lasts for 25 years from creation.

The stages in protecting your work by copyright are as follows.

1 Have the idea in the first place.

2 Turn it into something tangible by writing it, drawing it, creating the website, etc.

3 Date your work and apply the copyright © symbol with your name, and the date you created it.

4 Consider giving a copy of your work to a solicitor, or post it by 'recorded delivery' to yourself and leave it unopened, to provide evidence if you need to take action because someone who has infringed your copyright.

	Trademark	Copyright
What does it protect?	Brand identity (including words, logos and signs)	Music, art, literary works an broadcasts
What form does the protection take?	Prevents anyone using your trademark without permission	Prevents anyone copying or reproducing your work
Where does it operate?	Throughout the UK	Throughout the UK and most of the world
How long does it last?	Can last forever (must be renewed every 10 years)	Your lifetime +70 years (broadcasting and sound recording – 50 years)

Activity

Read through the copyright and trademark information for the Pocket PC shown, then answer these questions.

1 What is the date of the copyright and who owns it?

2 What is the status of the marks 'HP' and 'iPAQ'?

3 Describe the Secure Digital Logo.

4 Who owns the Bluetooth™ trademark?

Activity

Read each of the following statements and decide whether they are true or false.

1 Anyone can use the copyright © mark to protect a trademark.

2 Anyone can use the registered ® mark to protect a trademark.

3 Copyright is a private right in which you, the copyright owner, must decide how to exploit your copyright work.

4 Copyright must be registered to provide protection. You can do this easily but it will require annual renewal.

5 A copyright owner can decide whether or not to use the copyright work and/or license others to use it.

Just checking

✳ What is a trademark and what are the advantages of registering it?

✳ What is copyright and what does it apply to?

Research and development

Most companies don't leave the development of new products to chance. Instead, they invest considerable money in research and development, where new ideas are investigated and improvements are made to existing products and processes.

Research

Research is something that we all do when we need to find something out. We can do it systematically, with a particular goal or aim in mind, but sometimes we do it without a firm idea of what we will end up with. For example, suppose you are about to start a new job and this will involve you having to travel to work each day. You might research the public transport routes to see where they go, what they cost, and when they run. Alternatively, you might decide that the only sensible way of getting to work is to use your own transport. The problem then turns to researching what type of transport you will be able to afford and what it will cost to run in terms of loan repayment, insurance, fuel costs, etc.

Research is a crucial stage in producing a new product. It involves searching for new solutions, techniques or processes, while development involves finding ways in which these solutions can be put to practical use. Research is normally carried out ahead of development, but the two processes are often carried out side by side. You can think of research as a process that informs development and development as a process that is based on research. All of this means that you can't normally have development without research nor does it make sense to carry out research and not follow it up with some form of development.

There are two main forms of research:

* pure research
* applied research.

Pure research covers activities that add to scientific knowledge. It may, or may not, have any immediate commercial application. Applied research is directed towards specific commercial objectives. Also, many companies carry out market research to establish a specific market need and to ensure that the company only produces products and services that will actually sell. Market research is usually conducted by marketing specialists, whereas pure research is conducted by scientists and applied research largely by development engineers and technologists.

Development

Development is concerned with translating research findings into products and processes. A new product can take time to bring to market during which new materials, processes and techniques are introduced to help with its manufacture. It can often go through several stages of functional testing, and assessing whether it is economic to make and its market potential.

You should recall from Unit 2 that the design solution for a new product must meet the detailed requirements laid down in the design specification. This helps the development engineer by giving a very clear indication of how the product or service should perform.

Particularly important is the information on how the product will be made or assembled, and what materials and processes are to be used. The development process needs to provide definite answers to these questions, and also ensure that the product can be manufactured cost-effectively.

The Toyota Prius

Case study: The Toyota Prius

Takeshi Uchiyamada was born in 1946 and joined Toyota in 1969. It was always his dream to design a car but it was not until 1994, after years of work in software development and research into noise and vibration, that he finally got his chance. His brief was to develop a new car, which would be both fuel-efficient and environmentally friendly. He and his team quickly settled on a hybrid engine design, but they had to solve many technical and engineering problems before the car could be put into production.

At the time, advanced technologies like fuel cells and the electric vehicle were too expensive for a commercial product, so Uchiyamada initially proposed an advanced petrol engine. This idea was rejected because it lacked imagination and so the team set about working to a three-year deadline to produce the first hybrid petrol/electric car to enter production. To meet the demanding timescale, Uchiyamada abandoned the usual back-up plans and multiple scenarios found in a development project. Instead, he focussed his team on a single development goal, which was achieved one stage at a time. Finally, in 1997 the first Prius went on sale in Japan.

One of the biggest problems the team had to face was that the heavily used battery had to last for at least seven years. Research indicated that the battery would last longer if it was not fully charged and then fully discharged, as is usual with other types of battery. The team found that, by keeping the battery no more than 60% charged and no less than 40% discharged, its life was significantly extended. Among its other innovations, the Prius uses a more complex power train, together with a sophisticated **algorithm** that decides when to switch between the battery and petrol driven engines.

Looking back, the Prius was not the kind of car that most manufacturers would have ever approved of as a commercial project. The idea was simply too risky and there could have been too many unknown and insoluble problems along the way. If conventional thinking and standard decision-making had been followed, the car would have never been made. However, in this particular instance, conventional wisdom was wrong and Toyota's once sceptical rivals are now all busy developing their own hybrid vehicles.

Toyota has sold around 400,000 of the hybrid vehicles worldwide and in so doing has establishing an entirely new class of car. Takeshi Uchiyamada is currently an Executive Vice President at Toyota.

Activity

Read the case study for the Toyota Prius and answer the following questions.

1 What need did this product satisfy?

2 What entirely new class of vehicle was created?

3 Describe TWO innovative features of the Toyota Prius.

4 How did Takeshi Uchiyamada's approach to the development process differ from convention?

5 Why did he adopt this approach?

6 What did the Prius research reveal about the conditions under which a battery is used and the life that can be expected from it?

Just checking

✳ What are the main types of research and how do they differ?

✳ What is development and how does it relate to research?

Engineering companies need money to operate. This can come from sales of existing products and services, but that may not be enough to fund new product development. Fortunately, there are several ways of raising extra cash on both a short and long-term basis. They include **loans**, leasing and venture capital.

Borrowing

Engineering companies need money to pay their employees, for stock of materials, energy, transport, packaging, advertising, marketing and a host of other things. They also need to borrow money to develop and manufacture new products. Income from sales and services can be a major contributor to meeting these costs, but there's often a need for companies to raise extra finance.

Borrowing is often classed as short-term or long-term, depending on the period over which the loan is paid (i.e. the **repayment period**). A long-term loan might be used to finance the cost of a new building (an appreciable expense) while a short-term loan might be used to purchase a particular item of equipment. Long-term loans might be repaid over periods of typically between 12 and 25 years. Short-term loans are usually less than five years. Companies often borrow money on a short-term basis to acquire equipment or vehicles, or to obtain funds while they are in the process of arranging longer term finance.

Interest

Interest is the payment that a borrower makes in return for the use of the money borrowed. Loan repayments are often made on a monthly basis with the interest added. Interest rates are usually quoted as a percentage of the loan and they vary according to the prevailing economic climate. Typical interest rates vary from about 6% to around 10% per annum, depending upon the amount and period of the loan.

Example

A company wishes to borrow £120,000 and repay this over a period of five years. If the interest rate is 10% per annum, the monthly repayments would be calculated using an amortization table like this.

Over the six-year period, the total interest payable would be £33,000 and the monthly repayment would be £2,550.

Over the period of the loan, the original amount is repaid by regular payments. Each monthly payment includes part repayment of the original amount (the principal) plus an amount of interest. For example, at the end of the first month, the £2,550 payment will be split into two parts: £1,550 principal and £1,000 interest (one twelfth of 10% of the remaining balance). At the end of the second month, the £2,550 payment will comprise £1,560 principal and £990 interest (again, this is one twelfth of 10% of the remaining balance). At the end of the final (60th) month, payment will be £2,530 principal and £20 remaining interest.

Payback period

When borrowing money to fund new expenditure, most companies think in terms of the time that it would take for them to get their initial investment back. This is known as the payback period. Because of uncertainties about the future economic climate, most companies would want this to be as short as possible. Payback periods of between one and five years are not unusual.

Example

A company has decided to invest £200,000 in new plant and equipment. This is expected to produce additional income of £2,000 at the end of the first year, £30,000 at the end of the second year, £168,000 at the end of the third year, and £150,000 at the end of the fourth year. The initial cost (£200,000) would be paid back at the end of year 3. In year 4, the project would be returning a profit to the company.

Sources of finance

The most common source of finance is a clearing bank (often called high street banks), such as Barclays or HSBC. Banks usually lend fixed sums of money for fixed periods either at a fixed rate of interest or at an interest rate that is at the bank's base rate plus an agreed percentage.

Venture capital

You will know how venture capital works, if you have ever watched the TV programme 'Dragon's Den'! With venture capital, you may be able to gain financial investment to assist with the development of new products and processes in exchange for a share in your company (known as equity). Venture capital involves risk and those that loan money for this type of venture usually want a return on their investment within a fairly short period. There are lots of companies and individuals offering this sort of finance, but candidates for investment are rejected unless they are particularly attractive to lenders.

Leasing

A lease is a legal contract under which the owner of an asset (e.g. a building or vehicle) grants another person or company the right to use it for a specified period for an agreed fee. The owner is referred to as the lessor, whilst the user is the lessee. With a lease you do not own the equipment, you just have the right to use it!

Leasing is particularly valuable when a company wants to avoid tying up large capital sums for long periods of time. It is appropriate for business premises and vehicles, but can be applied to other capital equipment.

Loan – an amount of money given by a lender to a borrower

Repayment period – the time that it takes to repay a loan

Interest – the payment that a borrower makes in return for the use of the money borrowed

Payback period – the time that it takes to obtain a return on an investment

---

Activity

1 A company wishes to borrow £100,000 and repay it on a monthly basis over three years. If the annual interest rate is 7.5%, use a spreadsheet or other loan calculator to determine the total interest charged and monthly repayments.

2 A small company wishes to borrow £60,000 to fund the purchase of a new machine tool. The company can afford a monthly repayment of £1,750. If the simple interest rate is 10%, use a spreadsheet or other loan calculator to determine the minimum loan period that the company should choose.
Hint: You will find suitable software at www.key2study.com

Just checking

* What is interest and how is it calculated?
* Why is the difference between simple interest and APR?
* What is leasing and what is it used for?
* What is venture capital and what is it used for?

Materials and processes: metals

You will be familiar with metals such as aluminium, iron and copper in a wide variety of everyday applications, such as aluminium saucepans, copper water pipes and iron stoves. Metals can be mixed with other elements (often other metals) to form an **alloy**. Metal alloys are used to provide improved properties, because they are often stronger or tougher than the pure metals.

Classification of metals

You will already know about the main types of metal in use, such as iron, steel, copper and brass. Metals are often divided into two main classes: ferrous and non-ferrous, as shown in the chart. They can exist in both pure form and as alloys that are a mixture of several metals. For example, copper is used as pure material and combined in an alloy that we call brass. Metal alloys are used because they are often stronger or tougher than the pure metal. Other improvements can be made to metal alloys by heat-treating them as part of the manufacturing process. Thus steel is an alloy of iron and carbon and small quantities of other elements. If, after alloying, the steel is quickly cooled by quenching in oil or brine, a very hard steel can be produced.

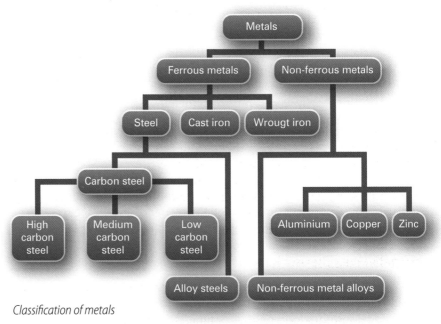

Classification of metals

Ferrous metals

Ferrous metals are based on iron. For engineering purposes, iron is usually associated with various amounts of the non-metal carbon. When the carbon present is less than 1.8%, we call the material steel. The figure of 1.8% is the theoretical maximum. However, in practice there is no advantage in increasing the amount of carbon present above about 1.4%, as shown in the graph.

Cast iron

Cast irons are also ferrous metals but have substantially more carbon than the plain carbon steels. Grey cast irons usually have between 3.2% and 3.5% carbon. Not all this carbon can be taken up by the iron and some is left over as flakes of graphite between the crystals of metal. It is these flakes of graphite that gives cast iron its particular properties and makes it a 'dirty' metal to machine.

Low-carbon steels

Low-carbon steels (also known as mild steels) are the cheapest and most widely used group of steels. Although the weakest, they are stronger than most non-ferrous metals and alloys. They can be hot- and cold-worked and machined with ease.

Medium-carbon steels

These are harder, tougher, stronger and more costly than low carbon steels. They are less ductile (meaning that they don't stretch easily) and cannot be bent or formed to any great extent in the cold condition without risk of cracking. Greater force is required to bend and form them. Medium-carbon steels hot-forge well, but close temperature control is essential. They divide into two ranges. The lower-carbon range can only be toughened by heating and quenching (cooling quickly by dipping in water). They cannot be hardened. The higher-carbon range can be hardened and tempered by heating and quenching.

High-carbon steels

These are harder, stronger and more costly than medium-carbon steels. They are also less tough. High-carbon steels are available as hot-rolled bars and forgings. Cold-drawn high-carbon steel wire (piano wire) is available in a limited range of sizes. Centre-less ground high-carbon steel rods (silver steel) are available in a wide range of diameters (inch and metric sizes) in lengths of 333 mm, 1 m and 2 m. High-carbon steels can only be bent cold to a limited extent before cracking. They are mostly used for making cutting tools, such as files, knives and other tools requiring a sharp edge. You will find more information on the properties of various types of carbon steel listed in the matSdata student materials database, which you can download free from www.key2study.com.

Non-ferrous metals

Non-ferrous metals (i.e. metals that are not based on iron) include metals such as aluminium and zinc as well as alloys such as brass and bronze. We shall start by looking at copper – a material that is widely used in plumbing and electrical engineering.

Copper

Pure copper is widely used for electrical conductors and other electrical components. It is second only to silver in conductivity, but it is much more plentiful and very much less costly. Pure copper is too soft and ductile for most mechanical applications. For general purpose applications such as roofing, chemical plant and ornamental work, tough-pitch copper is used. This contains some copper oxide which

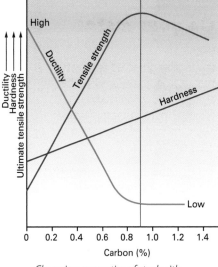

Changing properties of steel with carbon content

Round bar

Round tube

Square bar

Square tube

Rectangular tube

Rectangular bar

Hexagonal bar

Plate

Angle

Beam

Channel

Various forms in which ferrous and non-ferrous metals are supplied

makes it stronger, more rigid and less likely to tear when being machined. Because it is not so highly refined, it is less expensive than high conductivity copper. Copper is also the basis of many important alloys such as brass and bronze.

The general properties of copper are:

* relatively high strength
* very ductile so that it is usually cold-worked. An annealed (softened) copper wire can be stretched to nearly twice its length before it snaps
* corrosion-resistant
* second only to silver as a conductor of heat and electricity
* easily joined by soldering and brazing. For welding, a phosphorous deoxidised grade of copper must be used.

Copper is usually supplied as cold-drawn rods, wires and tubes. It is also available as cold-rolled sheet, strip and plate. Hot-worked copper is available as extruded sections and hot stampings. It can also be cast. Copper powders are used for making sintered components. It is one of the few pure metals of use to the engineer as a structural material.

Brass

Brass is an alloy of copper and zinc. The properties of a brass alloy and the applications for which you can use it depends on the amount of zinc present. Most brasses are attacked by sea water. The salt water eats away the zinc (known as dezincification) and leaves a weak, porous, spongy mass of copper. To prevent this, a small amount of tin is added to the alloy. Naval brass and Admiralty brass are two brass alloys that have properties that make them ideal for use at sea or on land near the sea.

Brass is a difficult metal to cast and brass castings tend to be coarse grained and porous. Brass depends upon hot rolling from cast ingots, followed by cold rolling or drawing to give it its mechanical strength. It can also be hot extruded and plumbing fittings are made by hot stamping. Brass machines to a better finish than copper as it is more rigid and less ductile than that metal. You will find the different types of brass and their properties listed in the matSdata student materials database which you can download for free from www.key2study.com.

Bronze

Bronzes are alloys of copper and tin. These alloys also have to have a deoxidising element present to prevent the tin from oxidising during casting and hot-working. If tin oxidises, the metal becomes hard and scratchy and is weakened. (This is the equivalent of 'rust' on oxidised ferrous metals.)The two deoxidising elements commonly used are:

* zinc in the gunmetal alloys
* phosphorus in the phosphor-bronze alloys.

Unlike brass alloys, the bronze alloys are usually used as castings. However low-tin content phosphor-bronze alloys can be extensively cold-worked. Bronze alloys are extremely resistant to corrosion and wear and are used for high-pressure valve bodies and heavy-duty bearings.

Aluminium

Aluminium has a density approximately one third that of steel. However, it is also very much weaker, so its strength/weight ratio is inferior. For stressed components, such as those found in aircraft, aluminium alloys have to be used. These can be as strong as steel and nearly as light as pure aluminium.

High purity aluminium is second only to copper as a conductor of heat and electricity. It is very difficult to join by welding or soldering and aluminium conductors are often terminated by crimping. Despite these difficulties, it is increasingly used for electrical conductors, where its light weight and low cost compared with copper is an advantage. Pure aluminium is resistant to normal atmospheric corrosion but is unsuitable for marine environments. It is available as wire, rod, cold-rolled sheet and extruded sections. The latter are useful for heat dissipators, used to keep electronic components cool.

Commercially pure aluminium contains up to 1% silicon to improve its strength and stiffness. As a result, it is not such a good conductor of electricity, nor so corrosion-resistant. It is available as wire, rod, cold-rolled sheet and extruded sections, but also as castings and forgings. Being stiffer than high purity aluminium, it machines better with less tendency to tear. It forms non-toxic oxides on its surface, which makes it suitable for food-processing plant and utensils. It is also used for forged and die-cast small machine parts. Because of their range and complexity, the light alloys based upon aluminium are beyond the scope of this unit.

Titanium

Titanium was discovered several hundred years ago but the technology for processing and using it has only been developed relatively recently. Titanium is light and very strong and, unlike many other metals, it is highly resistant to corrosion. Titanium has a relative density that is just over half that of steel, which makes it ideal for critical applications where a very high specific strength (or strength to weight ratio) is required. It also has a very high melting point (about 1,700° C) and so is ideal for use in high-temperature applications, such as gas turbine engines. Because of its resistance to corrosion, the metal has also found application in the petrochemical industry and also for surgical implants.

> **Metal** – chemical element (such as iron or copper), or mixture of elements (such as brass or bronze), made up of a large number of crystals. Metals are generally hard and strong, and are good conductors of heat and electricity
>
> **Alloy** – a metal that is made by mixing two or more metals

Many of the parts of this aircraft gas turbine engine are made from titanium

Just checking

* What are the main classes of metals?
* What are metal alloys and why are they useful?
* In what different forms are metals supplied?
* What is the effect of introducing carbon into steel?
* What are the advantages of titanium compared with steel?

Polymer materials or 'plastics' are not particularly new. In fact, the first semi-synthetic plastics appeared in the 1860s, and plastics made out of natural polymers have been used for centuries. Many other plastic materials were developed in the 20th century, and some were in mass production well before the Second World War.

Polymers

Polymers have the ability to be (initially at least) moulded into shape. They often have long and complicated chemical names. There is considerable incentive to seek more convenient names and abbreviations for everyday use. You might be familiar with PVC and PTFE but not polyvinyl chloride and polytetrafluoroethylene!

Polymers are made from molecules, which join together to form long chains in a process known as polymerisation. There are essentially three major types of polymer:

* **thermosetting plastics**, which once manufactured remain in their moulded form and cannot be re-worked

* **thermoplastics**, which have the ability to be remoulded and reheated after manufacture

* **elastomers or rubbers**, which have elastic properties (they can be stretched and will subsequently return to their original shape after stretching).

Thermosetting plastics

Themosetting plastics (also called thermosets) are available in powder or granular form and consist of a synthetic resin mixed with a filler. The filler reduces the cost and modifies the properties of the material. A colouring agent and a lubricant are also added. The lubricant helps the plasticised moulding material to flow into the fine detail of a mould.

The plastic is subjected to heat and pressure in the moulds during the moulding process. The hot moulds plasticise the moulding material so that it flows into all the detail of mould, and the heat also causes a chemical change in the material.

This chemical change is called polymerisation or, more simply, curing. Once cured, the moulding is hard and rigid. It can never again be softened by heating. If it is heated enough, it will just burn.

Thermosets include phenolic resins (e.g. Bakelite). These materials are hard, strong and rigid and are easily moulded and cured. Unfortunately they darken during processing and are only available in the darker and stronger colours. Phenolic resins are also used in the manufacture of plywood and laminated plastic materials, such as Tufnol. Polyster and epoxy resins (widely used as adhesives) are also examples of thermosetting plastic materials.

Thermoplastics

Unlike thermosets, thermoplastics soften every time they are heated. This makes it possible to easily recycle such materials, because waste materials can be ground up and recycled easily. Thermoplastics tend

---

Why do we use plastics?

Plastic materials of various sorts have become part of our everyday lives. Working with one or two other students, think of things that you use every day that are manufactured from plastics. Then think about whether they could have been manufactured using other materials, such as wood or metal. List four important features of plastics that make them attractive for use in everyday products.

Polymer – materials with molecules that join together to form long chains (called polymerisation).

Thermosetting plastic – polymer material that can be shaped and cured into a harder, more rigid form, through heat or chemical reaction. It cannot subsequently be softened by heating

Thermoplastic – polymer material that softens when heated and hardens to a rigid state when cooled. Further heating will cause the material to return to its softened state

Elastomer – polymer materials with elastic properties (they can be stretched and will return to their original shape). Elastomers are usually thermosets but can exist as thermoplastics

to be less rigid but tougher and more flexible than the thermosetting materials. Examples of thermoplastic materials include acrylics (e.g. Perspex or Plexiglass), polytetrafluoroethylene (PTFE), nylon, polyester (e.g. Terylene) and a variety of vinyl plastics including polythene, polypropylene, polystyrene and polyvinylchloride (PVC).

Another thermoplastic material, acrylonitrile butadiene styrene (ABS), is very commonly used see http://en.wikipedia.org/wiki/Thermoplastic to manufacture light, relatively rigid, moulded products, such as boxes, pipes, automotive body parts, wheel covers, enclosures and protective head gear. ABS parts are often black or grey in colour.

Elastomers

Elastomers are polymer materials with elastic properties. Common examples are plastic bands and car tyres. They are usually thermosets but can also be thermoplastics. Thermosetting elastomers like rubber require a process called vulcanising. This is a form of pressure moulding that is performed at a temperature of about 170°C for around 10 minutes.

Reinforced plastics

Plastics can be significantly strengthened by reinforcing them with fibrous materials. Examples include:

Laminated plastics – Fibrous material such as paper, woven cloth, woven glass fibre, etc. is impregnated with a thermosetting resin (e.g. Tufnol). Sheets of impregnated material are laid up in powerful hydraulic presses, then heated and squeezed until they become solid and rigid sheets, rods, tubes, etc. This material has a high strength and good electrical insulation properties. It can be machined with ordinary metal working tools and machines. Tufnol is used for making insulators, gears and bearing bushes.

Glass reinforced plastics (GRP) – Woven glass fibre and chopped strand mat can be bonded together by polyester or epoxy resins to form mouldings. These may range from simple objects, such as crash helmets, to complex hulls for ocean-going racing yachts.

The thermosetting plastics used with GRP are cured by chemical action at room temperature and a press is not required. The glass fibre is laid up over plaster or wooden moulds and coated with the resin, which is well worked into the reinforcing material. Several layers or plies may be built up, according to the strength required.

When cured, the moulding is removed from the mould. The mould is coated with a release agent before moulding commences, so that it can be used again. We shall return to this topic in the next section when we describe composite materials in more detail.

General properties of plastic materials

Although the properties of plastic materials can vary widely, they all have some general properties in common including excellent strength/weight ratio, excellent corrosion-resistance and good insulation properties.

Activity

Brief descriptions of five polymer materials are as follows:

Material A: Supplied in sheets and in various colours; a good electrical insulator with excellent thermal forming properties and good impact strength; ideal for use in the manufacture of cases and enclosures.

Material B: A rather expensive material; has very low friction and adhesion properties; suitable for use over a very wide temperature range; can be extruded or supplied as a coating; has excellent electrical properties.

Material C: A thermoplastic material with good sliding properties; easily bonded, welded and machined; resistant to oils and grease; suitable for use in the manufacture of gears, pulleys and machine parts.

Material D: A rather weak material supplied in opaque clear sheets or as a film; has very high electrical resistance and is ideal for use as an insulator.

Material E: Supplied in clear sheets; a very rigid material; ideal for use as a cover for light fittings; windows and door panels.

(a) Identify materials A to E by matching them to the following list: nylon, ABS, polycarbonate, PTFE, polythene.

(b) Which of the materials would be suitable for the manufacture of (i) a roller bearing, (ii) a damp-proof membrane, and (iii) a 'non-stick' coating.

Just checking

* What is a polymer material and what properties does it have?
* What are the main classes of polymer material and how do they differ?
* How can the properties of a polymer material be improved?
* What processes are used with polymer materials?

Polymers perform well when used as electrical insulators at normal temperatures. But when things get really hot, they might melt and cause a short-circuit. Suppose you are dealing with something that gets really hot – like the interior of a furnace. Working with one or two other students, think of at least three other situations in which plastic materials (such as PVC) can't be used as electrical insulators due to the temperatures involved. What material is used in each of these situations and why?

Materials and processes: ceramics and composite materials

In addition to metals and polymers, several other classes of material are used in engineered products, including ceramics and composite materials. Both have properties that make them particularly useful in certain applications.

Ceramics

Ceramics are chemical compounds formed from oxides such as silica (sand), sodium and calcium, as well as silicates such as clay. They have very high melting temperatures and, on their own, are difficult to form. They are usually made more manageable by mixing them, in powder form, with water, and then hardened by heating.

Ceramics include:

* a wide variety of glass products
* abrasive and cutting tool materials
* construction industry materials, such as cements and plasters
* electrical insulators
* refractory (i.e. heat resistant) lining for furnaces
* heat resistant coatings for metals.

The four main groups of ceramics are amorphous, crystalline, bonded and cements. Here are the properties of ceramic materials.

* **Strength** – Ceramic materials are reasonably strong in compression, but tend to be weak in tension and shear. They are brittle and lack ductility. They also suffer from micro-cracks, which occur during the firing process. These lead to fatigue failure. Many ceramics retain their high compressive strength at very high temperatures.

* **Hardness** – Most ceramic materials are harder than other engineering materials. Crystalline ceramics (such as silicon carbide) are used as abrasives and cutting tool materials. They retain their hardness at very high temperatures that would destroy high-carbon and high-speed steels. However they have to be handled carefully because of their brittleness.

* **Refractoriness** – This is the ability of a material to withstand high temperatures without softening and deforming under normal service conditions. Some refractory materials, such as high-alumina brick and fireclays, tend to soften gradually and may collapse at temperatures well below their fusion (melting) temperatures. Refractory materials made from clays, containing a high proportion of silica to alumina, are most widely used for furnace linings.

* **Electrical properties** – As well as being used for weather-resistant high-voltage insulators for overhead cables and sub-station equipment, ceramics are now being used for low-loss high-frequency insulators – for example, dielectric in silvered ceramic capacitors for high-frequency applications.

(a) Aligned continuous

(b) Aligned discontinuous

(c) Random discontinuous

Different fibre alignments in a composite material

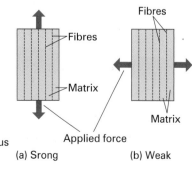

Fibres

Fibres

Matrix

Matrix

Applied force

(a) Srong

(b) Weak

Construction of fibre mats

Composite materials

Composite materials are quite different from any of the materials that we have met so far. As their name suggests, composite materials are combinations of materials that retain their separate identities and do not dissolve or merge together.

In structural applications, composite materials have these characteristics.

✱ They generally consist of two or more physically distinct and mechanically separable materials.

✱ They are made by mixing the separate materials in such a way as to achieve controlled and uniform dispersion of the constituents.

✱ The mechanical properties of the composite material are superior to (and in many cases quite different from) the properties of the constituent materials.

Composite materials are ideal for manufacturing aircraft parts with complex aerodynamic profiles

A good example of a composite material is glass-reinforced plastic (GRP). This material combines glass fibres with epoxy resin. The latter material is relatively weak and brittle and, although the glass fibres are strong and stiff, they can only be effectively loaded in tension as single fibres. However, when combined into a composite material, the resin and fibre provide us with a strong, stiff material with excellent toughness. Various materials are commonly used as fibres including alumina, boron, carbon, glass, polyethylene and polyamide. Matrix materials include alumina, aluminium, epoxy, polyester and polypropylene.

The fibres may be arranged within the matrix material in various ways depending on the properties required and the intended application. Note that, whenever the fibres are aligned in a specific direction, the properties of the material also become directional. Conversely, if the fibres are arranged in random orientation, the resulting material will have the same properties in all directions. This is an important point, because we may sometimes require that a component has maximum strength in a particular direction. In this case, the correct alignment of the fibres becomes extremely important.

(a) Chopped strand mat

(b) Continuous filament mat

(c) Bi-directional woven mat

Effect of applying a force to a material with aligned fibresferrous metals

Ceramic – a non-metallic material that is formed by the action of heat

Composite material – a material made from two or more constituent materials that retain their separate identities and do not dissolve or merge together

Personal learning and thinking skills

Find out about the construction and operation of a conventional spark plug. Prepare and deliver a ten-minute presentation with the aid of appropriate visual aids to the rest of your class. Identify the materials used and say why they are used. Also, explain how the temperature range for a spark plug is determined.

Just checking

✱ What is a ceramic and what are the properties of this type of material?

✱ What is a composite material and why are they used?

✱ What is glass reinforced plastic (GRP) and what gives it its properties?

Optoelectronics – application of electronic devices that produce or respond to light

Data communication – transmission and reception of digital information (usually over some distance)

Engineering and new technology

The use of new technology is an expanding part of engineering. New technology is used in all sectors of engineering and the speed at which new developments are introduced is set to gain pace for the foreseeable future. This is an exciting time to be an engineer!

Applications of new technology in engineering

So far in this unit, you have seen the need for innovation in engineering and the stages that engineering designers go through when developing new products. You have also discovered some of the range of materials and processes available to the engineer. In the next few pages, you will look briefly at how new technology has been used in several branches of engineering, including:

* the use of new materials and coatings in aerospace
* the use of electronic control systems in the automotive sector
* **optoelectronic** devices in consumer electronics
* the use of fibre optics in **data communications**.

Materials and coatings in the aerospace industry

The aerospace industry need to design and manufacture aircraft to exacting standards, within budget and on time. At the same time, we expect our aircraft to perform better, fly longer and further, use less fuel and have a reduced impact on the environment. A prime consideration in aircraft design is the materials from which it is made. Construction methods must then guarantee the structural integrity of the aircraft when subject to the stresses that it encounters in operation. Companies such as Airbus (EADS), British Aerospace (BAe), Westland, Rolls-Royce and Boeing have responded to this challenge with new materials and coatings used in a wide range of civil and military applications.

Example: The Airbus A380

The Airbus A380 is the largest passenger aircraft in the world. To minimise weight, the A380 incorporates a variety of new materials, including:

* carbon-fibre reinforced plastic – central wing box, horizontal stabilisers, fin, rear fuselage section, ceiling beams
* glare – new, very light material, saving about 500 kg on panels for the upper fuselage. Aluminium and fibreglass layers do not allow cracks to spread
* aluminium-lithium – used as a skin on parts of the fuselage and also in the wings of the freighter version of the aircraft, the A380F. Stronger than traditional materials, which means a thinner coat can be applied to the fuselage. This results in further weight-savings without loss in strength

The A380F will be the first freighter to use a carbon-fibre barrier between the cockpit and aircraft forward section and cargo area.

Lighter than traditional aluminium, it is also corrosion-resistant, reducing maintenance.

The Airbus A380

Automotive electronics

Electronic systems are widely used to improve engine efficiency and vehicle performance. They include electronic control units (ECU) and complete engine management systems (EMS) which are found in more sophisticated vehicles. The ECU is a computerised system that takes inputs from different sensors and provides outputs to control fuel delivery and ignition. Other electronic systems on a modern vehicle may include anti-lock braking systems (ABS), traction control systems (TCS), ride control systems (RCS), air conditioning systems (AC), driver information systems (DI), and in-car entertainment systems (ICE).

Example: Engine management systems (EMS)

EMS are becoming increasingly powerful and sophisticated to improve fuel efficiency and reduce unwanted emissions. They can provide control over most aspects of engine performance and incorporate a number of sub-systems, such as those used to control fuel injection and ignition.

Most EMS use microprocessor-based systems to measure engine speed and the load imposed on it. The EMS uses a look-up table of injector opening times (known as the base fuel map) stored in non-volatile memory (i.e. memory that retains data when the power is off). The data held in the base fuel map is for a specific engine type but it is usually possible to change this using external programming software.

Having determined engine requirements from the base fuel map, the microprocessor system can perform adjustments and correct for air temperature and pressure (which affect the density of air used in the combustion process). Extra injection time can be added for sudden throttle movement and engine temperature. The calculations provide the final injection time (the time for which the injectors are held open).

Injection pulses usually occur one or more times per engine cycle. The ECU uses a trigger signal synchronised with the engine cam (and therefore locked to engine speed) to give the precise time to inject fuel into the cylinders. The ECU then applies a pulse of magnetising current to the injector coils for the computed injection time. This ensures extremely accurate fuel delivery, exactly matching engine needs and operational circumstances.

Example: Automatic diagnostic systems

Being able to deal effectively with a variety of electronic systems is increasingly challenging for those involved with motor vehicle maintenance. The success of any workshop is increasingly dependent on its ability to diagnose and repair faults in vehicle systems and the components (such as the igniters) that they control.

Many vehicles now incorporate dedicated diagnostic connectors into which sophisticated external diagnostic equipment can be plugged. More compact, hand-held diagnostic testers can be used to make an initial evaluation of engine performance and determine the nature of a fault before more detailed investigation in the workshop.

Activity

So-called 'smart' or 'responsive' materials are becoming increasingly important. These respond to changes in their environment by changing their properties (mechanical, electrical, optical, appearance), structure or composition, or their functions. They include shape-memory alloys (SMA); metals that, after being strained, will at a certain temperature revert back to their original shape. Using your library or Internet resources, investigate SMA use in either (a) actuators and control systems or (b) medical applications. Present your findings as an A4 factsheet.

A typical engine management system showing the ECU

Activity

A company called DuPont invented Mylar polyester film in the early 1950s. Investigate the development and use of Mylar in the manufacture of electronic components. Write a brief word-processed report giving your findings.

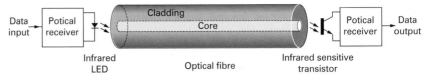

| Data input | Potical receiver | | Cladding / Core | | Potical receiver | Data output |

Infrared LED   Optical fibre   Infrared sensitive transistor

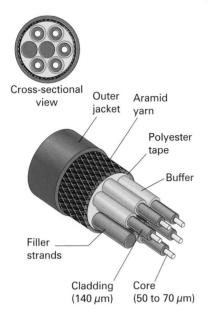

Cross-sectional view

Outer jacket — Aramid yarn

Polyester tape

Buffer

Filler strands

Cladding (140 µm)   Core (50 to 70 µm)

Typical optical fibre construction

Consumer electronics

Developments in semiconductor technology have led to an increasingly wide range of sophisticated devices for use in consumer electronic equipment. Three of the most notable areas of development have been:

* an expanding range of consumer integrated circuits
* increasingly larger semiconductor memory devices
* optoelectronic devices, indicators and displays.

Example: Light emitting diodes

One of the earliest optical components to become widely available was the light emitting diode (LED). Compared to filament lamps, the LED provides vastly superior reliability and significantly improved efficiency.

LEDs comprise a small slice of semiconductor material that forms a P-N junction (see Unit 5). As with a conventional diode, the current in an LED flows from anode (the P-type material) to the cathode (the N-type material). Inside the LED, charge carriers (electrons and holes – see Unit 5) flow across the junction when the diode is forward biased and conducting. When an electron meets a hole, it falls into a lower energy level and releases energy in the form of a photon of light.

Most LEDs produce light at the red end of the visible spectrum. Others produce infrared light (beyond the visible range) and are used for remote controls and data communications, using optical fibres to conduct the emitted light from an optical transmitter to an optical receiver. Other colours are also possible, including yellow, green and blue.

Example: CDs and DVDs

Compact discs (CDs) have been around for nearly 20 years. They can store 65 minutes of high-quality recorded audio or around 700 Mbytes of computer data, roughly equivalent to 250,000 pages of A4 text. Introduced in the early 1990s, CDs quickly established themselves in computing and hi-fi sectors and are still widely used.

CDs have three layers. The main part is an injection-moulded polycarbonate substrate, which incorporates a spiral track. This has pits separated by lands (i.e. raised areas between the pits) cut into it corresponding to the data to be stored on the disk. Over the substrate is a thin aluminium (or gold) reflective layer, which in turn is covered by an outer protective lacquer coating. Information is retrieved from the CD by focusing a low-power infrared laser beam onto the spiral track. The optoelectronic system senses the difference in the height of the pits and

Human hair diameter = 64 µm

40 tracks

Recording tracks on a CD compared with a human hair

lands and produces a digital signal (a series of '1's and '0's) from the recorded data.

The digital versatile disk (DVD) first appeared in the late 1990s. It uses the same video technology as digital TV, satellite and cable transmissions. A full-length movie can fit onto one side of a DVD at roughly broadcast TV quality, plus multi-channel digital sound. The film production companies immediately saw it as a way of stimulating the video market, producing better sound and pictures at considerably less cost than VHS tape, and requiring a much simpler player than the complex mechanical tape transport system used on VHS equipment.

For computer users, the DVD offered much greater data storage on a single disk and it was largely through sale of computer DVD-ROM drives that the DVD market took off. In the late 1990s, computer-based DVD drives outsold DVD players by more than 5:1, largely due to original equipment manufacturers (OEM) incorporating DVD drives into new computer equipment.

By improving manufacturing tolerances and the optical playback system, a seven-fold increase in data capacity can be achieved over the CD.

✱ Tracks are closer together, allowing more tracks per disc (DVD track pitch is less than half that of a CD).

✱ Data storage pits are much smaller, allowing more per track (DVD pit length is less than half that of a CD).

✱ DVDs are available single and double sided (CDs are only single), and support single or dual layer recording (an additional semi-transparent layer, also containing data, enabling two layers of data on one side).

Depending on the type of DVD, and whether it is single- or double-sided, and single- or dual-layer, disk storage capacities range from around 4.7 GB to a massive 17 GB.

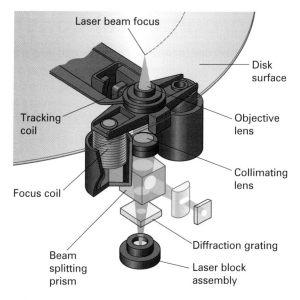

Laser beam focus

Disk surface

Tracking coil

Objective lens

Focus coil

Collimating lens

Beam splitting prism

Diffraction grating

Laser block assembly

The optical unit of a CD player produces a precisely focused beam of laser energy

Activity

Blu-Ray is the latest technology for storing video information. Use library or internet resources to locate information on Blu-Ray technology.

1 What major development in TV and video has made it necessary to develop Blu-Ray technology?

2 A number of well-known companies are supporting the development of Blu-Ray technology. Name FIVE of them.

3 What makes it possible for Blu-Ray disks to store more data than conventional DVD?

Activity

Use your library or internet resources to answer the following questions.

1 DVD technology has made VHS equipment obsolete. Why?

2 Why was the movie industry keen to replace the VHS recorder?

3 What technology was used for high-quality sound recording in the 1950s and 1960s? How did it work and what were its main limitations?

The optical unit of a DVD recorder

Just checking

✱ Why is new technology important in engineering?

✱ What does an engine management system do?

✱ What does an engine diagnostic system do?

✱ What are optoelectronics?

✱ What is data communication?

Environmental constraints

We have become increasingly concerned about the impact of industries like coal, steel and petroleum. All require large plant, which generates noise and pollution. Smaller engineering activities also affect the environment. For example, producing car parts brings more traffic onto the roads. This puts pressure on local resources and can be unpleasant for those living and working nearby.

Engineering and the environment

Many engineering activities involve large-scale processing of materials. Some, such as wood and air, occur naturally and may require only minimal treatment before use. Others, like steel and ceramic materials, must be manufactured and processed from raw materials.

Extraction of raw material, its processing or manufacture, creates wealth and social benefits. For example, extraction of iron-ore in Cleveland and processing into pure iron and steel has attracted people to live in the area. This benefits local shops and entertainment centres, local builders must provide more homes and schools and local services improve. Arguably, there is a better quality of life.

However, extraction of raw materials can leave the landscape untidy. Slag heaps and disused quarries are not a pretty sight. In recent years, much effort has been expended on improving these eyesores. Slag heaps have been remodelled as part of golf courses, and disused quarries filled for water sports or fishing. Transporting raw materials from place to place also causes noise and pollution.

The effects of waste products

Engineering activities are a major source of **pollutants**. Air, soil, rivers, lakes and seas can be polluted by waste gases, liquids and solids discarded by industry. Heat can also be a pollutant.

Electricity is a common source of energy and its generation often involves burning fossil fuels: coal, oil and natural gas. Each year, billions of tonnes of carbon dioxide, sulfur dioxide, smoke and toxic metals are released into the air. Some electricity-generating stations use nuclear fuel, which produces highly radioactive solid waste. Exhaust gases from motor vehicles, oil refineries, chemical works and industrial furnaces are also sources of toxic or biologically damaging pollutants. Non-toxic waste, like plastic and metal scrap dumped on waste tips, slag heaps and around old industrial sites, also pollutes the environment.

Pollutants are defined as degradable or non-degradable depending on whether the pollutant decomposes or disperses with time. Smoke is degradable, but most dumped plastic waste is not.

* **Carbon dioxide (CO_2)** – CO_2 in the air absorbs some of the long-wave radiation emitted by the earth's surface and, in so doing, is heated. The more CO_2 there is in the air, the greater the heating or 'greenhouse effect'. This is thought to be a major cause of global warming, causing average seasonal temperatures to rise. The increased quantity of CO_2 in the air, especially around large cities, may also cause respiratory problems. We have become concerned about the '**carbon footprint**' of human activities – the amount of CO_2 generated or released by an activity or process.

Not in my back yard!

Working with one or two other students, decide on which of the following engineering plants you would least like to have built in your neighbourhood: steel-rolling mill; car production plant; oil refinery; nuclear power station or wind farm. Give reasons for your answer.

Pollutant – an unwanted by-product of an engineering process (such as noise, heat or smoke)

Carbon footprint – the amount of carbon dioxide generated or released by an activity or process

Ozone layer – a naturally occurring layer in the upper atmosphere that protects us from harmful solar radiation

An example of industrial pollution

* **Oxides of nitrogen** – Oxides of nitrogen are produced in most exhaust gases and nitric oxide is prevalent near industrial furnaces. Fortunately, most oxides of nitrogen are soon washed out of the air by rain. But if there is no rain, the air becomes increasingly polluted and unpleasant.

* **Sulfur dioxide** – Sulfur dioxide is produced by burning fuels that contain sulfur, such as coal. High concentrations of this gas cause peoples' lungs to constrict and breathing becomes difficult. It also combines with rain droplets to form sulfuric acid or acid rain, which can fall many hundreds of miles from the source. Acid rain increases the normal weathering effect on buildings and soil, corrodes metals and textiles and damages vegetation.

* **Smoke** – Smoke is caused by incomplete burning of fossil fuels. It is a health hazard on its own but even more dangerous if combined with fog. This poisonous combination, called smog, was prevalent in major cities in the early 1950s, when many deaths were recorded. This led to the first Clean Air Act, which prohibited the burning of fuels that caused smoke in areas of high population. So-called smokeless zones were established.

* **Dust and grit** – Dust and grit (or ash) are very fine particles of solid material formed by combustion and other industrial processes. The lighter particles may be held in the air for many hours. They form a mist, which produces a weak, hazy sunshine and less light.

* **Toxic metals** – Toxic metals, such as lead and mercury, are released into the air by some engineering processes and by exhaust gases. They can be carried hundreds of miles before falling in rainwater to contaminate soil and vegetation. Motorists are encouraged to use lead-free petrol to reduce the level of lead pollution.

* **Ozone** – Ozone is a gas that exists naturally in the earth's stratosphere where it helps reduce harmful ultra-violet radiation from the sun. In the 1980s, it was discovered that gases from engineering activities were causing a 'hole' in the **ozone layer**, which may increase the risk of skin cancer, eye cataracts, and damage to crops and marine life. At ground level, sunlight reacts with motor vehicle exhaust gases to produce ozone, which causes breathing difficulties. This 'tropospheric' ozone is a key constituent of 'photochemical smog' or summer smog. In the UK, it has increased by about 60% in the last 40 years.

* **Heat** – Heat is a waste product of many engineering activities, such as hot coolant water from electricity-generating stations, which can destroy aquatic life. This waste is reducing because of measures to improve energy efficiency.

* **Chemical waste** – Chemical waste dumped directly into rivers and the sea, or onto land near water, can cause serious pollution. There is also long-term danger from chemicals dumped on soil soaking into the ground water, which we then drink.

* **Radioactive waste** – Radioactive waste from nuclear power stations, or other engineering activities, poses particular problems. Not only is it a powerful cause of cancer but it remains dangerous for scores of years. Present methods of disposing of radioactive waste include their encasement in lead and burial underground or at sea, but this remains contentious.

* **Derelict land** – Land can be so badly damaged by, for example, mining or quarrying, that it cannot be used without treatment or restoration work.

Activity

Find out what happens to the domestic waste produced in your locality. What items of domestic waste are recycled? Is any of the waste burnt to produce useful heat? If the waste is transported to another site for processing, where does it go and what processes are used? What arrangements are there for disposing of hazardous waste? Present your findings in the form of a brief report. Illustrate your report using a flowchart, which shows the sequence of processing that is applied to different types of waste.

Activity

Give three examples of air pollution caused by engineering activities.

Activity

Explain how damage to the ozone layer has resulted in an increase in the incidence of skin cancer.

Just checking

* What is a pollutant?
* Give four examples of pollutants.
* What is a carbon footprint?

How has it affected you?

Engineering has had an impact on all of us. Working with one or two other students, think of at least five things that you do (such as air travel) that would not be possible without engineering. Then list these in order of importance to you. Which would you be least prepared to do without and why?

Energy-efficiency – a measure of how much energy is wasted by a product. In order to reduce wasted heat and reduce a product's 'carbon footprint' it is necessary to ensure that energy efficiency is as high as possible

Sustainable – property of a product that uses raw materials that can be regenerated or re-grown. Wood is a potentially sustainable material, because forests can be replanted when trees are cut down to produce paper and building materials. Petroleum based products are unsustainable products because they will eventually run out

Recycling – reuse of a product when no longer needed. Recycling usually involves extraction of materials such as metals, plastics and glass so that they can be re-used in the manufacture of new (and possibly different) products

Biodegradable – property of a product that becomes less of a waste hazard due to biological action. Paper is biodegradable, while metals and plastics are not

To complete this unit, you will need to find out about the environmental and social aspects of engineering and be able to apply these factors to an engineering product. The assessment for this unit is based on a case study of an engineering product or service, so this unit comes to a conclusion by looking at some of the questions that you might need to answer.

The social impact of engineering

There can be little doubt that engineering has affected the way we live. In this final section, you look at a number of ways in which engineering has had a direct impact on us.

Television

The domestic television set has had a huge impact on society. From 1945, the TV began to replace sound-only radio as the main domestic source for daily news and information. The original national TV programmes were transmitted from local sites, only by the BBC and only in black and white. The programmes – mainly news reports, sports commentaries or shows – were not broadcast until 7 pm and closed down at midnight. Watching the evening TV programme had the importance of a family visit to the local cinema. In the 1950s, as 7 pm approached, families that could afford a TV receiver (many couldn't!) would be riveted to their 12-inch (30 cm) TV screen, awaiting the single-channel 405-line picture that would arrive courtesy of the BBC.

Today, we have an ever-increasing choice of programmes to watch. They are broadcast by the BBC and independent (ITV) television companies using chains of land-based transmitters and geostationary satellites. Many programmes are now received live from around the world using satellite relays. In addition, cable companies are providing domestic cable subscribers with a range of cable TV programmes. The 625-line pictures that we now enjoy are invariably clearer, in colour and broadcast 24 hours a day. There are screen sizes to suit any room – no longer are we restricted to a mere 12 inches (30 cm) or 14 inches (36 cm)!

With such a huge range of TV programmes available on a continuous basis, it's not surprising that viewing audiences are attracted at all hours of the day and night. The viewing pattern in many homes, especially those with young children, is for the TV to be switched on at breakfast time and left running until bedtime. The programmes cover everything from high-quality educational programmes and current affairs to family entertainment comprising news, sport, comedy and drama, together with a contentious amount of sex, horror and violence. Also available is a vast amount of travel, weather, business, statistical and other written information on text-only channels, Ceefax (BBC) and Teletext (ITV). Last, but by no means least, are the advertisements shown at approximately 15-minute intervals on all ITV channels. These are a major source of revenue for the independent TV companies and income from them finances the programmes that we enjoy.

Educational or straightforward entertainment programmes, undoubtedly have a beneficial effect upon society. They can be informative, thought-provoking, amusing or just relaxing. However, the continual aural and visual battering to ourselves and others is blamed for many shortcomings in society today. Young people, in particular, are said to be very much influenced by the increasing amount of violence seen on TV; and, unfortunately, young viewers may begin to regard violent behaviour as a normal part of life.

Refrigerators and freezers

Refrigerators and freezers are now part of our everyday lives and virtually every household has both. Before widespread use of refrigerators, fresh food had a very limited life, especially in summer. Milk and meats could be kept fresh in a cool cupboard for one or two days only, which meant that fresh supplies were essential. The refrigerator, with its internal temperature a little above (+4°C) the freezing point of water (0°C) and its hermetically sealed door, has at least doubled the storage time of fresh food, and keeps rodents and insects away.

Two immediate benefits have been a healthier range of fresh food readily available in the kitchen, and the need to go shopping reduced to perhaps once or twice a week. On the negative side, the refrigerator costs money to buy and to run. Freezers store food at an even lower temperature (typically −18°C) and the food remains in a solid frozen state. The benefits are even longer food storage times – often between three and six months – coupled with the economy of bulk buying of frozen meats and ready-cooked meals. Freezers also allow us to enjoy food out of season, such as peas and beans in the winter, long after they have disappeared from our gardens.

Unfortunately, many older refrigerators and freezers contain chlorofluorocarbons (CFC) and/or hydrochlorofluorocarbons (HCFC) either in the refrigerant of the cooling system or as a blowing agent in the insulating foam. These materials are classed as ozone-depleting substances (ODS) and must be disposed of responsibly. Most manufacturers stopped using CFCs as a refrigerant in the mid-1990s, but HCFCs continued to be used as a blowing agent for insulating foam for several years after that. For this reason, the majority of refrigerators require special waste disposal treatment and users must take responsibility for their correct disposal.

Mobile phones

Mobile phones let us keep in touch with friends, family and work colleagues wherever we are. They are a relatively recent invention, initially only covering major cities and with phones too bulky and expensive for most people. Nowadays they are tiny and inexpensive, and there are few places where a network signal is unavailable.

The instant communication offered by mobile phones has saved many lives and minimised the impact of natural disasters. However, it is dangerous to use a mobile phone when driving and this has now been recognised in law. They can also be annoying and distracting when used inconsiderately and there are also concerns about users' exposure to electromagnetic radiation. Conclusive evidence to support this has yet to be presented but doubts still exist about the safety of long-term exposure.

Activity

Most households own a car and many have several. The car has many advantages, the main one being its convenience as means of personal transport. Check the number of cars belonging to several of your immediate neighbours. What is the average number of cars per household, what are they used for and roughly how many miles do they travel each year? Compare your results with those obtained by another student living in a different area. Discuss the reasons for any differences.

Personal learning and thinking skills

TV receivers and video recorders are often fitted with a 'standby mode'. In this mode, part of the circuit continues to operate, and therefore consumes power, when the equipment is not in use. Use a library or the Internet to explore the pros and cons of using consumer electronic equipment on standby. Present your findings as a brief article for publication in a local newspaper. Make sure the article is suitable for non-technical readers.

Personal learning and thinking skills

What negative impact has the motor car had on society and what alternatives are there to personal car ownership? Present your findings to the class using appropriate visual aids. Your presentation should last no more than ten minutes and you should allow a further five minutes for questions.

Personal learning and thinking skills

A magazine for elderly readers has commissioned you to write a short article (no more than 1,500 words) describing specialist medical equipment likely to be available in a local hospital. It is to be suitable for non-technical readers and will appear as part of a feature on how new technology affects the elderly. You should aim to describe several of the most important items of equipment and explain, in simple terms, what each is used for and why it benefits older people.

Cars

Widespread ownership of cars has had a huge impact in all developed countries. They enable people to travel to work using their own transport rather than public transport (e.g. bus or train). This has led to the demand for improved roads and motorways carrying an increasingly heavy volume of traffic. Cars also take up space when they are not being used. Urban roads and streets are now full of parked cars where no other provision for parking exists. Finding a parking space has become a regular challenge for many drivers.

Cars generate pollution from pollutants that appear in their exhaust gasses and also due to noise. This is particularly unpleasant for those living close to a major road or motorway. In some cases, pollutants in exhaust gasses can be sufficient to form a health hazard. Large clouds of exhaust gasses fill the air above most major cities when there is no breeze.

Most car manufacturers are investing considerable resources to develop engines that are both **energy-efficient** and burn fuels that generate less pollution. Catalytic converters are now fitted to most cars to reduce emissions, and hybrid cars are being introduced by a number of manufacturers in an attempt to become 'greener' and reduce the 'carbon footprint'. In addition, schemes are being trialled in various parts of the country to reduce the number of cars on the road. These include car sharing and lane restrictions on cars with single occupancy.

Medical electronics

Engineered products are increasingly important in health care. There is a wide range of sophisticated medical diagnostic equipment, such as magnetic resonance imaging (MRI) and computerised tomography (CAT) scanners, which combine engineering and computer technologies. Some of these advanced techniques explore not only the structure of part of a person's body but also whether it is functioning correctly. They may examine the electromagnetic and mechanical properties of the atoms of an organ within the body, looking at sectional slices. A computer can then analyse the electrical characteristics of the atoms in each slice and then construct an image of the whole organ.

Employment

Engineering advancements have, in many cases, had a significant impact on the need for large numbers of unskilled and semi-skilled workers. The manually operated, moving conveyor-belt production line, which passes a product along a line of stationary workers with each performing some simple task, is dying out.

In the main, this low-skilled work has been taken over by automated production lines. Typical of these are the robot-operated, motor car body assembly lines. Also, in microelectronics, the manufacture of integrated circuits, and their subsequent assembly onto printed circuit boards, is so detailed and fine that this is possible only by using computer-controlled machines.

New technology is also bringing about a big change in the type of worker required. The concept of each worker having well-defined work to perform – a job description – is, in some companies, regarded as counter-productive. They argue that it can lead to a 'that's not my job' restrictive approach, where work is passed around and doesn't get done. A new approach is team work, to see the whole project completed properly and on time. While an individual may have a special skill that the team requires, if he or she has other skills, then they should be put to good use. In most jobs, the idea of a worker with a single skill has ended. Instead, we require workers to have several skills and so become multi-skilled. This gives greater flexibility, increased job satisfaction and improved job security.

Activity

What is meant by a 405-line TV picture? Why is a 625-line picture of higher quality? See if you can find out how many lines actually appear on the screen (not all of them do!). What is the difference between HDTV and normal 625-line TV?

Just checking

* What is an ozone depleting substance and why should they be avoided?
* What does biodegradable mean?
* What is energy efficiency and why is it important?
* What does sustainable mean?
* What is multiskilling and why is it important?

Unit 8 Assessment Guide

In this unit, you will investigate how engineers develop good ideas into products and services that people will buy and use. Invention and development can be costly processes, and businesses have to raise finance in order to fund them. Businesses also need to be able to protect their ideas from competitors.

Time Management

Manage your time well as this unit has a number of different components that need to be researched. Ensure that any notes you take are kept safe and work held in electronic format has a secure and safe backup.

The unit is externally assessed by an examination which lasts 90 minutes, and you will be given the opportunity to practice for it.

The examination is based on a case study of an engineering product or service – this will be given to you prior to the examination so that you can conduct research. You will need to manage your time effectively as you carry out this preparation.

Be well organised. This is your chance to show that you are an independent enquirer, creative thinker, reflective learner and self manager and therefore will contribute towards achievement of your Personal learning and thinking skills.

Be prepared with a list of relevant questions that you could ask if your teacher arranges a visiting speaker or a visit to an engineering company.

Plan ahead for your work experience and make a list of things that you need to find out so that you make the most efficient use of your time.

Useful Links

Make good use of your work experience and arrange to talk with people who work in the design, development and finance departments of the company.

A visit to a museum or exhibition will be helpful when investigating the impact that engineering technologies have had in the home, workplace and built environment.

This website will give you information about intellectual property and patents: http://www.patent.gov.uk

You will need to access a materials database such as www.matweb.com

Things you might need

You would benefit from access to a professional engineer working within the design sector of the industry.

Remember to maintain a focus how innovation in engineering is linked with advances in technology.

A trial/revision examination paper would be useful.

Examples of how you could be assessed in the examination

What you must show that you know	Guidance	To gain higher marks
How intellectual property within engineering can be used by companies to protect their inventions. *Assessment focus 8.1*	✱ Describe the four main types of intellectual property.	✱ You must explain the benefits of registering intellectual property. ✱ You must explain the problems an inventor has to overcome when trying to protect their rights (e.g. the Dyson/ Hoover courtroom battle)

That research, development and raising finance all have an important role when a company designs a new product.
 	
Assessment focus 8.2
	✱ Describe the role, sources and significance of research and development when designing a product or process.
	✱ Describe the following methods which a small business could use to finance a project:
	  ✱ bank loan
	  ✱ grant
	  ✱ private funding.
		✱ You must describe and review the way that an engineering project is financed (you need to research a real project).
		✱ You must explain how publicity can be used to gain financial backing for a new product: e.g. Trevor Bayliss and his clockwork radio.

Why developments in materials and processing have had an effect on the design of products.
 	
Assessment focus 8.3
	✱ Produce a timeline which charts the development and use of titanium alloys over the last 50 years.
	✱ Describe how a named material from each of the following categories is used in a "hi-tech" way:
	  ✱ metal
	  ✱ polymer
	  ✱ ceramic
	  ✱ composite.
		✱ You must explain why a design engineer will take account of the following factors when choosing a material for a particular application:
		  ✱ properties
		  ✱ characteristics
		  ✱ costs
		  ✱ forms of supply
		  ✱ availability.
		You must select an engineered product and explain why the designer chose the material from which it was made: e.g. why is stainless steel used for the inner shell of a dish washer?

How engineering technologies have affected the home, workplace and built environment.
 	
Assessment focus 8.4
	✱ Produce a time line which charts the development of personal music systems.
	✱ For each of the following applications identify 4 products which are "new technology" and describe their purpose:
	  ✱ in your home
	  ✱ in an engineering factory
	  ✱ in a hospital
	  ✱ in a high rise office block.
		✱ You must review the effect that advances in technology have had in the home, workplace and built environment; e.g. improvements in living and working conditions, better medical care, and more efficient use of resources.

That there is an environmental and social impact associated with engineering and a requirement to consider sustainability of resources.
 	
Assessment focus 8.5
	✱ Identify examples of the impact that engineering has had on:
	  ✱ the environment
	  ✱ the way we live in the 21st Century.
	✱ Choose a product and explain how the following factors influenced the choice of material from which it was made:
	  ✱ recycling potential
	  ✱ reusability
	  ✱ bio-degradability
	  ✱ sustainability.
		✱ You must review the environmental and social impacts of engineering: e.g. are engineers killing or saving the planet?

Your project

Introduction

Welcome to one of the most important parts of the course – your own, unique project!

As you know, the Level 1 and Level 2 project can be completed as a stand-alone qualification or as part of your Diploma. You can choose any topic that fits with the Diploma, and will have 60 hours to complete it, with the support of your project tutor or assessor at every stage.

Why are you being asked to do a project? Because it gives you the opportunity to extend your learning in a topic or subject that you find particularly interesting or useful. You can explore your chosen subject or topic in depth, and produce a piece of work that is unlike anyone else's.

Your project will be assessed against the following criteria.

1 Manage your project – for this you will need to:

* complete a project proposal form
* explain why you have chosen your topic, and describe the skills and knowledge that you want to improve
* identify your project objectives
* plan your project activities and agree deadlines
* identify possible risks and how to overcome them
* keep records of all of your activities.

2 Use of resources – for this you will need to:

* research your topic using a range of different resources and different types of resource
* evaluate the reliability of your sources
* keep records of all relevant information that you find.

3 Develop and realise your project – for this you will need to:

* complete your project
* achieve your project objectives
* share your findings.

4 Review your project – for this you will need to:

* analyse your findings
* draw your own conclusions
* review your performance.

There are three different types of project outcome: ephemeral, artefact/design and written. This table shows a few examples of the type of evidence that could be presented for each.

Type of evidence	Example of evidence
Ephemeral	A performance or one-off event, such as a role play or group activity. The outcome should be recorded, probably on video, and must show evidence of the stages that you have gone through, and how your ideas have been developed.
Artefact or design	You will need to provide: a description of the problem you wish to solve; sketches, diagrams or drawings of the design; and explanations of how your artefact works. You will also need to provide evidence of improvements in your design throughout its development.
Written	A written report with findings and recommendations. As a minimum, your report will need to show what the project is about, what you have done and what your findings were.

To carry out your project successfully, you will need to do a number of things, including:

* choosing your project topic
* planning your project
* researching your project
* presenting your completed project
* evaluating your project and your own learning.

Here you will learn about all these aspects of your project.

BEFORE YOU START

Find yourself a notebook that you can use to write down everything related to your project. This will become your project notebook, and will form valuable evidence of successful project management and research activity.

Choosing your project topic

The first stage of your project is choosing a topic. Take some time to think about the topics that you have studied as part of your course, or other topics that you have heard about that interest you. It is a good idea to list a few different topics as initial ideas, which you can discuss with your project tutor before deciding which one to do.

The box below shows the principle learning areas for your Diploma and may give you some ideas.

Level 1 Topics	Level 2 Topics
Introducing the engineering world	Exploring the engineering world
Practical engineering and communication skills	Investigating engineering design
Introduction to computer aided design	Engineering applications of computers
Developing routine maintenance skills	Producing engineering solutions
Introduction to engineering materials	Electrical and electronic circuits and systems
Electronic circuit construction and testing	Application of manufacturing techniques in engineering
Engineering the future	Application of maintenance techniques in engineering
	Exploring engineering innovation, enterprise and technological advancments

Some dos and don'ts for choosing your project

Do

✓ Make a list of several topics that interest you

✓ Talk to your project tutor about your possible topics – this will help you to make the right choice

✓ Choose a topic that you are really interested in

✓ Choose a topic that is related to your Diploma

Don't

✗ Choose a topic because you know somebody else is doing it

✗ Start your project without talking to your project tutor

✗ Choose a topic that is not related to your Diploma

Planning your project

Once you have identified your project topic, produce a project brief or a topic-specific question that identifies your **aims** and **objectives**. For example, if you choose to carry out your project on the topic of Engineering design, your project brief should provide you with the opportunity to fully explore different types of Engineering design programs and their possible uses. Look at the example below and think how you might produce your own project brief.

Writing aims and objectives

Take some time to think about your own personal aims and the specific objectives that will help you to achieve your aims. For example, think about where you would like to be in five years' time and write it down in your project notebook. This is your aim. Next, think about the things that you will need to do, or achieve, in order to get there. These will be your objectives. Remember to make sure that your objectives are SMART. If you are not sure whether or not they are SMART, consider asking your project tutor to check them for you. The case study below may help you with this activity.

Case Study: Chloe's report

Chloe is a product designer who is working for a telecommunications company that designs components for transferring data. The company has manufacturing premises worldwide and employees who speak numerous languages. Due to this, the design information sent between sites must be clear and concise. All relevant employees must be able to understand the requirements of the designs that are being sent electronically. One of the manufacturing sites has highlighted that at times there are problems with the files: this could be due to the type of design file created. Chloe has been requested by her line manager to investigate the use of engineering design packages used within the company. Chloe's aim is to produce a detailed report on engineering design programs.

Her **objectives** are:

1 Make a list of different types of engineering design programs that are available.

2 Highlight the additional features that the engineering design programs come with, including an explanation for these additional features.

3 Recommend the most suitable feature for testing the materials of a component that has been designed.

4 List all of the common file extensions that the design can be saved as, and explain the features of these.

5 Describe why one of the engineering sites may be experiencing a problem with opening some of the designs.

6 Produce a recommendation based on this research of the most desirable design package.

What are aims and objectives?

The **aim** is to produce a report on engineering design programs. The **objectives** are the individual tasks or targets that the project manager will need to complete in order to achieve the aim.

Objectives must be **SMART**:

✳ **S**pecific – an objective must contain specific detail about what needs to be done

✳ **M**easurable – you must be able to measure progress towards achievement and recognise completion

✳ **A**chievable – your objectives must be achievable within the constraints that you have

✳ **R**ealistic – you must actually be able to achieve your objectives

✳ **T**ime-bound – there must be a time limit for achieving your objectives.

Tips for writing objectives

✳ Remember to make sure your objectives are SMART.

✳ Keep your objectives as simple as possible – you need to know when each objective has been achieved.

✳ Try to order your objectives in some kind of logical sequence – it may help you to stay focused if you know that you cannot complete an objective until the one before is finished.

✳ Don't have too many objectives – you may end up feeling that you will never complete the project.

Aim – a broad statement of intent: for example, 'In this project I aim to explore different types of structure and their different uses'

Objective – what you are actually going to do: for example, 'List three different types of structure and state their most common uses'. It may help you to think of objectives as targets

Bibliography – a list of magazines, books, websites and newspapers that have been used in your research

Hyperlink – an active link in a document that takes you directly to a web address when you click on it

Preliminaries – information that comes before the main part of a report

Milestone – a point at which you will demonstrate progress towards your outcomes

How should you manage your time?

Once you have chosen your topic and written your project proposal form, you will need to start thinking about how long each part of the project is going to take. Think about how long you have to complete the project and then set yourself some deadlines for completing various stages. This information must be included on your project proposal form and you will have two **milestone** dates that must be agreed with your project tutor or assessor. For example, if you have 12 weeks to complete your project, you might set yourself a time limit of one week to choose your topic and discuss it with your project tutor. Then your next time limit might be for deciding your aims and objectives.

You should map out the entire project with dates for achieving certain things. These could be your milestones. This will help you to manage your project effectively and ensure that you meet your overall deadline for completion.

Example: setting deadlines

Your Diploma teacher is planning a visit to a nearby engineering company at the end of the term, which is in five weeks' time. To ensure that the visit goes ahead without any problems, they will need to plan it in plenty of time. Here are some examples of the things that they will need to think about and do during the five weeks leading up to the visit.

* Contact the engineering training manager to discuss the number of visitors, the arrival and departure times of the visit, the health and safety requirements, and any other significant issues that need to be addressed before the visit can take place. Deadline for task ~ Week 1.

* Confirm the number of students and supporting adults that will be attending the visit, book the transport and write the risk assessment to ensure that insurance is provided. Deadline for tasks ~ Week 2.

* Make sure students and supporting adults are briefed in detail about the visit and what they are looking for to enable them to complete the post-visit assignment that will be set. Deadline for tasks ~ Week 3.

* Discuss health and safety issues that are relevant to the company, and remind everybody of the importance of following instructions while on site. Deadline ~ week 4.

* Remind everybody of meeting time and place to ensure a prompt departure, and of all the resources that will be required, such as notebook and pen, questionnaire or company research. Deadline ~ Day before the visit.

Effective time management is vital if you are to complete your project on time. This means that, as well as setting yourself deadlines for completing the various stages or activities, you must also make sure that you identify the time itself. For example, you may decide to work on the project for three hours every week. If so, you will need to identify when those three hours will be and protect that time by making sure that nothing else interferes with it. Perhaps you will plan to do an hour on Monday, an hour on Wednesday and an hour on Thursday. Alternatively, you may decide it is better to do three hours every Sunday afternoon. Whatever you decide, it is important that you do not allow anything else to take that time. Tell the people that place other demands on you, like friends, family, etc. that this time is already taken and that you cannot spend it with them.

What structure should your project have?

Having written your aims and objectives, you can start to think about how the project will take shape.

If you are writing a report, or for the written part of your project, a good way to ensure that you include everything that you need is to use a writing frame. The following diagram shows a simple academic writing frame you could follow.

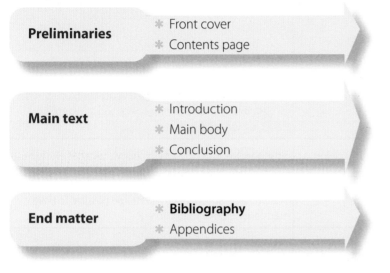

Preliminaries	* Front cover * Contents page
Main text	* Introduction * Main body * Conclusion
End matter	* **Bibliography** * Appendices

Preliminaries When completing a large piece of work, it is normal to include a front cover and a contents page. Your front cover will normally include information like your name, the name of the person you are doing the work for, the title of the project, etc. You should check what is required with your project tutor. The contents page will list everything that is included in your project, with page numbers for each section.

Main text The main text of your project will probably be split into three sections: introduction, main body and conclusion. The introduction should tell your reader what the piece of work is about and provide information on how you will tackle the question. It will also add context by providing some general information about the topic you are addressing. The main body of the project report is where you will answer the specific questions that formed the basis of your learning outcomes. Each paragraph should tackle a particular issue and each will lead into the next so that the report shows a logical development. The conclusion is where you will bring all of your ideas and findings together, and explain how each of your learning outcomes has been met. The conclusion also provides an opportunity for you to make recommendations based on your findings.

End matter After the conclusion, you should include a bibliography and any appendices. Make sure that you include all of the reference sources that you have used for researching your project in your bibliography.

Researching your project

When researching your project, you will need to think about where you can get the information that you need. There are thousands of books, journals and websites that you can access, but which ones will help you the most? And how can you find them? This table shows a range of different information sources you can try.

Top tip

Try setting yourself a rough word count for each section of the project. This will help make sure that you do not write too much on one part, or too little on another – and you'll be able to see the progress you are making!

Information source	Example
Books	Textbooks and dictionaries may prove very useful when completing your project. You could talk to your tutor about which books will help you most. Why not create a reading list in your project notebook?
Professional journals	Journals like Professional Engineer, TEP, (Technology Enhancement Program), Design Engineer and Engineering Times, all linked specifically to the Engineering industry, may be helpful. There are many other engineering journals that deal with specific areas of the industry, including processes, control, materials and systems. You will need to look around to find the ones most closely linked with the subject of your project. Your project tutor could help with advice and guidance on other relevant publications
Audiovisual resources	DVDS, CDs and various software packages may be useful – but don't think you have to buy them. Visit your library and see what is available for you to borrow: if resource is not available at your library, it may be possible to have it transferred from another library.
The internet	The internet is a huge resource that is easily accessible and often much quicker than searching through books and journals. However, don't try to do everything on the internet: using a variety of sources will give a much better, more rounded result – and the assessors will see it!

Books, journals and many audiovisual resources can be accessed at your local library, or your school or college library. All of the resources at a library are catalogued and can be found quite quickly. Libraries usually have a librarian who can help you to understand the referencing system and find the resources you need as quickly as possible. The internet, however, is not referenced in the same way and you may need to use a search engine to find what you are looking for. When using a search engine, you can enter keywords and it will find a huge range of sites where the keywords appear. However, not all of them will be relevant and you will need to work through them to find the ones that are most useful. Unlike published books or journal articles, information on the internet does not always go through an editing process and anybody can create their own site and put information on it. This means that not everything on the internet is accurate or up to date. In some cases, information on the internet that is presented as fact is not even true.

Another factor to consider when researching your project is how far back you wish to look and how many different sources of information you will use. If, for example, your project relates to something like manufacturing or engineering processes, you need to find the most recent information possible because over the years, with the improvement of technology, systems and processes improve. This means that information from just two years ago could well be out of date, and modern or smarter processes are now being used.

Although books, journals and the internet will all prove to be useful resources while researching your project, you may also wish to consider other research methods, such as interviews or questionnaires. The opinions and views of people who have knowledge and experience in the area that you are researching can be very useful. For example, if you are researching smart identification systems used in the engineering industry, you could interview line managers from local engineering companies. If there are a lot of engineering companies in your area, you could consider sending out a questionnaire. This is a much more efficient way of contacting large numbers of people. However, you need to remember that not everybody will respond to a questionnaire so think carefully before deciding to use one, particularly if you need responses from everybody.

Throughout the research stage of your project, remember to make use of your project notebook. Keep a record of all the information sources that you use. List book titles, authors, publishers, dates, etc. as this information will be needed for your bibliography. Don't forget to list web addresses, details about journal articles, and people that you have interviewed or spoken to informally about your project. If you have interviewed individuals, or groups, keep a detailed record of what was said.

Presenting your completed project

Having completed the research for your project, you now need to think about how you will present your project, state your findings and make recommendations. You may be completing a written report, producing an artefact or preparing a role play or other activity. If you are presenting your project as a written report, without any kind of presentation, remember to plan your writing using a writing frame and set yourself some boundaries for how many words will be in each section.

The use of PowerPoint can greatly enhance a verbal presentation. PowerPoint is a presentation program within the Microsoft Office suite that can be used to prepare slideshows and presentations, including visual effects, sound effects, charts, graphs, images and hyperlinks to help you get your point across.

Written	* You could present your project as a written report. Look back at the earlier section on using a writing frame to help you structure your report

Verbal	* This would involve making a verbal presentation to your project tutor and possibly a wider audience including your assessors and maybe your peers. You will probably need to present a written report to your tutor to support the verbal presentation.

Audio visual	* Much like a verbal presentation but using audio visual aids like video or PowerPoint to assist you. As with verbal presentation, you may still need to present a written report for sassessment purposes

Some PowerPoint dos and don'ts

Do

✓ Practise your presentation before you do it for real
✓ Make sure that you use a colour scheme or background that anyone can read
✓ Keep your presentation short and to the point
✓ Check that any **hyperlinks** in your presentation are still live
✓ Know what is on each slide and talk about it, without reading it word for word

Don't

✗ Put too much text on each slide
✗ Use more slides than you need
✗ Make your presentation too long
✗ Read from your slides

Evaluating your project and your own learning

After all the hard work of choosing, planning, researching and presenting your project, you now need to review your project and evaluate your learning. Spend some time looking through your project notebook and reflecting on the various stages of your project. Think about what things you have learned and what skills you have developed along the way. Maybe it is the first time that you have researched a topic on your own. Perhaps it is the first time that you have interviewed people or written a questionnaire to help you collect information. Remember, your project assessor will be checking that you have effectively reviewed your project and evaluated your own performance. Think about what type of project evidence you are presenting and how you could most effectively show that you have evaluated the project. Don't forget that, if you have done some kind of presentation, role play or other performance, feedback from those who observed it could provide useful evidence of how well you performed. Below, are some examples of questions that you could ask yourself as part of the review process

Project evaluation

* How many of your project outcomes did you meet?
* If any of your project outcomes were not met, why not?
* What conclusions have you drawn from your project?
* What problems did you encounter and how did you overcome them?
* What new skills have you gained from completing the project?
* Which of these new skills could be transferred to another area of work?
* If you were doing your project again, what things would you do differently?

Remember to thank everyone who has helped with your research. These people might be interested in reading/watching what you have found out, so be sure to keep them informed of the final result.

Congratulations!

The Diploma Project is a major piece of work. If you have worked through this section of the book and completed your project, you have done really well. In doing your project, you will have learned some very useful skills that could be of help to you in the future, even if you change your mind about your future career. Here's what successful completion of your project means you have gained:

* Decision-making skills – you thought about different topics before choosing, and made choices about how to carry out your project.

* Target-setting skills – you thought about your project aim and set yourself some learning objectives or targets.

* Planning skills – you planned your project and how you would do it.

* Time-management skills – you have learned about how to make the best use of your time to ensure that deadlines are met.

* Research skills – you have researched your project and found different sources of information.

* Academic writing skills – you have learned how to set out your project report using an academic writing frame.

* Presentation skills – as well as presenting a written report, you may have delivered a formal presentation to support your project findings.

* Evaluation skills – you have evaluated different sources of information throughout your project and evaluated your findings in order to make recommendations. Also, you have evaluated your own learning and performance against your project aims and objectives.

All of these skills have great value in everyday life and will be of use to you whatever career you eventually choose. Well done!

19th Century – 1800 to 1899 3-axes machines – CNC milling machines with relative tool and work piece movement in three directions (X, Y and Z axes)

Achievement – a milestone in history Activity log – a document that summarises the work carried out in order to arrive at a final design solution

Algorithm – a series of defined steps, or actions, that are followed to achieve a desired outcome

Alkaline storage battery – a type of battery or cell, dependant upon the reaction between zinc and manganese

Alloy – a metal that is made by mixing two or more metals

Alternating current (a.c.) – current that continuously changes direction, flowing first one way then the other. UK mains supply changes direction fifty times a second (a frequency of fifty cycles per second, or 50 Hz).

Analogue instrument – an instrument on which measurements are shown physically, such as a pointer on a dial

Analogue signal – a signal where the sound is reproduced as a radio signal, which varies in amplitude (size) to match the change in sound

Analyse – to investigate and explore to determine a result

Annotated – labelled, written information on an image

Anode – the more positive terminal of a diode when conducting

Attributes chart – used when a 'yes' or 'no' decision is being taken, such as a plug gauge

Automated – carried out partially or completely by a computer or machine

Automatic right – IP right not requiring a formal application or payment of a fee

Axes – the directions the tool is able to move in, usually by reference to three axes at right angles: X, Y and Z

Bar code – a numerical code that can be printed on objects, able to be read by a scanner and interpreted by a computer

Batch production – a quantity of engineered products manufactured in one run, that will need to be repeated if further quantities are required

Biodegradable – property of a product that becomes less of a waste hazard due to biological action. Paper is biodegradable, while metals and plastics are not

Bipolar junction transistor (BJT) – a transistor with two junctions (either NPN or PNP) and operates with both positive and negative charge carriers

Bistable circuit – a digital circuit that retains its logic state indefinitely or until set or reset. A bistable circuit acts like a simple memory, storing a logic 0 or logic 1 condition

Bluetooth™ – wireless short-range communication system

Body language – communication without words (e.g. smiles, movements, eye contact, etc.)

Brainstorming – a means of generating ideas from rapid suggestions made by a group of people (usually without spending a lot of time thinking in depth about the problem)

Burn-in – the operation of a product prior to delivery to a customer, intended to identify and reject early failures due to defective components or faulty manufacture

Candidate solutions – one of several potential design solutions to a particular design problem. The final design solution is selected from these

Carbon footprint – the amount of carbon dioxide generated or released by an activity or process

Cathode – the more negative terminal of a diode when conducting

CE mark – a means of identifying products that comply with the relevant European Directives (and can therefore be legally offered for sale in any EU state)

Ceramic – a non-metallic material that is formed by the action of heat

Chairperson – the person appointed to lead a meeting, who can steer the team

Charge – a negative charge results from an accumulation of electrons; a positive charge results from a lack of electrons.

Chip – very thin slice of silicon a few millimetres across, chemically etched with transistors and other components

Chip breaker – an aid to control the flow of swarf-chips from a cutting action in a continuous flow

Chip formation – how swarf is generated from a cutting action between tool and work piece

Chuck – a device that holds a drill or screwdriver bit in a rotary tool such as a drill. Chucks usually have three jaws and can be adjusted for a range of different bit sizes

Closed question – when a factual type of answer is required

Clutch – a mechanism that causes rotation (turning) when engaged and no rotation (no turning) when not engaged

CMOS – a logic family based on complementary metal oxide semiconductor technology

CNC program – a set of instructions that the CNC machine will convert into actions

Colour code – a system of coloured bands to indicate the value and tolerance of a component

Commercial product – a product that meets a market need at an acceptable price

Common emitter amplifier – BJT used as an amplifier with emitter connection common to both input and output

Compliance – being able to demonstrate that a set of prescribed criteria are met (there may be several ways of demonstrating compliance with a particular directive)

Components – parts that go together to make a product

Composite – a mixture of materials, put together to achieve certain characteristics

Computer numerical control (CNC) – a machine that can be reprogrammed and controlled by computer to complete a sequence of operations

Computer simulation – a computer model of an object or process

Concurrent operations – more than one activity carried out at the same time

Conductor – a material that conducts electric current. Metals are good conductors

Consumer – the end-user of a product (a customer who purchases and then uses a product is a consumer)

Corrosion – chemical attack on metals and metal alloy (for example, rust)

CPU (central processing unit) – the heart of a computer that carries out computation and processing

Cross-sectional view – a view showing the inside of a product as if it has been cut and opened out

Current – the organised movement of charge carriers (electrons in metal conductors)

Cutting speed – usually quoted in metres per minute, but the machine has to be set at revs per minute, so be careful when using this terminology

Database – a collection of data arranged for ease and speed of search and retrieval

Data communication – transmission and reception of digital information (usually over some distance)

Datum point – a point that work and tooling movements are referenced to

Degradation – gradual weakening of non-metals due to chemical attack, exposure to sunlight or other radiation

Design brief – a statement that identifies what is needed to satisfy a need or solve a design problem

Design criteria – a list of factors that need to be taken into account in a design

Design process – the various stages that are followed when preparing a design

Design proposal – a detailed solution to a design problem. Usually a range of different design proposals are considered before a decision is reached on which is best

Design solution – the answer to a particular design problem, usually presented in the form of a design proposal

Design specification – a set of detailed requirements that must be satisfied by the design solution

Detail drawing – a drawing that provides all of the information required to manufacture a component part

Detection – identifying that something is present or absent

Development – an action, invention or service that aids in making an achievement

Dielectric – an insulating material that separates the plates of a capacitor

Digital instrument – an instrument that takes a physical measurement, converts it electronically and displays it as a number

Digital integrated circuit – integrated circuit designed for digital applications

Digital multi-range meter (DMM) – a multi-range meter with a digital (usually LCD) display

Digital signal – a signal made up of 1s and 0s, as in binary code

Dimension – a measurement in respect to the size or feature on an engineered product

Diode – a semiconductor device that allows current to flow in one direction only (from anode to cathode)

Direct current (d.c.) – current that flows in one direction. Conventional current flow is from positive to negative. Electrons flow in the opposite direction!

Directive – an official EU document that defines a set of requirements

Dynamo – another name for an electrical generator, a machine that converts mechanical into electrical energy

Early failure – a failure that occurs during or shortly after manufacture, usually caused by faulty materials or manufacturing processes

Elastomer – polymer materials with elastic properties (they can be stretched and will return to their original shape). Elastomers are usually thermosets but can exist as thermoplastics

Electrolytic capacitor – a capacitor that relies on chemical action for its operation

EMC directive – the EU requirement for electromagnetic emission and susceptibility

Employee – a person employed by an employer under a contract of employment

Employer – a person or organisation that employs one or more persons under a contract of employment

Energy – the ability to do work. Energy efficiency – a measure of how much energy is wasted by a product. In order to reduce wasted heat and reduce a product's 'carbon footprint' it is necessary to ensure that energy efficiency is as high as possible

Engine control unit (ECU) – electronic system that controls the operation of several aspects of an internal combustion engine

Evaluation matrix – a chart with check marks or numerical scores used to compare different solutions against a list of criteria

Explicit knowledge – knowledge that exists on paper or in electronic form Exploded view – a pictorial representation of a product showing how the individual parts fit together

Failure rate – the total number of failures within a population, divided by the total number of life units represented by that population, during a particular measurement interval under stated conditions

Fastening – a means of securely joining two parts, such as a panel and its supporting frame or a machined component to its mounting block

Fault-finding chart – a diagram showing logical steps to find a fault on an electronic circuit quickly and easily

Feed rate – the rate at which the tool moves across or into the material as it is taking a cut, quoted in metres per minute or, more likely, mm per rev

Ferromagnet – a material (such as iron, steel, or ferrite) that supports a magnetic flux

Fixture – a work-holding device that positions the work piece in the same place each time

Flip chart – a set of A2 or A3 sheets that can be flipped over to reveal their contents. Flip charts are quick and simple to use and are ideal for use with smaller audiences

Forming shapes – relies on the tool shape to produce a feature or shape on the work piece

Forward voltage – the threshold voltage that needs to be applied to a diode for it to conduct (about 0.2 V for a germanium diode and 0.6 V for a silicon diode)

G-code – a computer language for writing code to control CNC machines

Gantt chart – a representation of a schedule of operations or activities along a time series

Gauge – equipment that helps decide if a dimension is within tolerance but without declaring actual size

General arrangement drawing – a drawing that shows a complete assembly and usually includes an itemised list of the parts used

Generating shapes – relies on tool and work movement to produce a shape

Hazard – a condition that

Hybrid integrated circuit – integrated circuit that uses both analogue and digital technology

Ignition – in vehicles, causing a fuel/air mixture to burn

Implicit knowledge – knowledge we hold in our minds but is not written down

Industrial Revolution – rapid development during the 19th Century, which saw a major shift of people from agriculture and traditional crafts to towns and factories

Inductor – a component that stores energy in the form of a magnetic field

Input – whatever goes into a computer, from commands entered from the keyboard to data from another computer device

Inspection – examination by sight, touch, smell and sound (as appropriate) to determine the functional state of a product, process plant or equipment

Insulator – a material that does not conduct electric current. Glass, ceramics and plastics are good insulators

Integrated circuit – semiconductor device with a large number of components on a single chip of silicon

Intellectual property (IP) – property created by intellectual or creative activity, including patents, trademarks, copyright and designs. Like all property, if you own it you can rent it, lease it, license it, give it away or sell it

Interest – the payment that a borrower makes in return for the use of the money borrowed

Knowledge-based organisation – an organisation that uses a fund of knowledge for its day-to-day operation.

Lead time – the time it takes from ordering the component to it arriving

Level of risk – the likelihood that an event will occur that also takes into account the severity of its outcome. If something is unlikely to happen but would have severe consequences it can still represent a relatively high level of risk. Similarly, if something is very likely to happen but will have only minor consequences it could be regarded as a high risk activity in which it is performed

Linear integrated circuit – integrated circuit designed for analogue (linear) applications

Loan – an amount of money given by a lender to a borrower

Logic gate – a digital circuit that performs a logic function, such as AND, OR, NAND and NOR

Low-Voltage Directive – the EU requirements for equipment normally operating at from 50 V to 1,000 V a.c. or 75 V to 1,500 V d.c

Machinability – how easily a chip can be formed and separated

Magnetron tube – electronic device used to generate microwaves

Maintenance – activity involved in keeping something in working order

Maintenance procedure – the logical sequence of tasks that should be performed when carrying out routine maintenance

Marking out – the process of preparing a component for manufacture by applying surface marks that correspond to features of the finished component (for example, the position of holes

Materials – what material the product is made from

Matrix board construction – a method of constructing an electronic circuit using a circuit board that has strips of copper tracks with holes arranged in the form of a matrix

Maximum stock level – the number of parts that need to be held in stock given the rate of use and the supplier's lead time. Replacement parts must be ordered in quantities equal to the maximum stock level

Mean chart – tracks the average size of components over time

Mean time to failure (MTTF) – the total operating time of a number of items divided by the total number of failures

Mechanism – arrangement of mechanical components that perform a function, such as opening a window

Metal – chemical element (such as iron or copper), or mixture of elements (such as brass or bronze), made up of a large number of crystals. Metals are generally hard and strong, and are good conductors of heat and electricity

Microprocessor – a computer on a chip

Microsoft PowerPoint presentation – a presentation using a software package that forms part of the Microsoft Office suite. It can be delivered to a large audience using a screen and projector connected to a computer

Microwave – a form of electromagnetic energy used in microwave cookers and other items like mobile phones

Milling – material removal process used to produce surfaces and shapes with the tool normally rotating

Mind mapping – a means of generating ideas using a diagram (or mind map), which shows different routes that can be taken to solve a problem

MRO – maintenance, repair and overhaul

Multi-range meter – an instrument with several measurement functions (voltage, current and resistance) and ranges

N-type – semiconductor material doped with an impurity to produce an excess of electrons in the lattice of a pure semiconductor material (such as silicon)

Negative feedback – measurement of a change in a process, comparing it with the desired value and using this to reverse the change back towards the desired value

Non-profit making organisation – an organisation in which profits are not distributed to its directors, shareholders, employees or anyone else. Instead, profits are reinvested into the services provided

Normal distribution – where the proportion of the measurements taken that lies between the mean and a specified number of deviations away from that mean is always the same

NVQ – National Vocational Qualification

OEM – original equipment Manufacturer

Open question – a type of question that will lead to debate and discussion

Operational amplifier – a type of integrated circuit designed for analogue (linear) applications

Optoelectronics – application of electronic devices that produce or respond to light

Oscilloscope – an instrument that provides displays of voltage against time (i.e. waveforms)

Output – anything that comes out of a computer. Output can be meaningful information or encrypted code and appear as binary numbers, characters, pictures, printed pages, etc.

Overhead projector – a standalone projector that uses A4 transparencies or rolls of acetate film. They can be used with large or small audiences, but can sometimes be crude and hard to read

Ozone layer – a naturally occurring layer in the upper atmosphere that protects us from harmful solar radiation

P-type – semiconductor material doped with an impurity to produce an excess of holes (i.e. gaps into which electrons can fit) in the lattice of a pure semiconductor material (such as silicon)

Patent – exclusive right to use an invention commercially in return for disclosing it and payment of fees

Patent application – the documents you need to file, giving details of your invention

Payback period – the time that it takes to obtain a return on an investment

PEO – Performing Engineering Operations, an industry related qualification

Personal protective equipment (PPE) – clothing, footwear and other items that can be worn or carried to avoid risks and minimise hazards

Pictogram – a picture that conveys meaning and can be widely understood. Pictograms are often used as an alternative to text, particularly where many different languages may be used

Polarising voltage – a d.c. voltage that needs to be applied to an electrolytic capacitor for it to operate

Pollutant – an unwanted by-product of an engineering process (such as noise, heat or smoke)

Polymer – materials with molecules that join together to form long chains (called polymerisation).

Potential diff erence – the voltage drop that appears across a circuit or component when an electromotive force (e.g. a battery) is connected across it

Potentiometer – a variable resistor with three terminals, two with fixed resistance and one (the slider) that is varied between the other two

Post-16 – over the age of 16

Power – the rate at which energy is converted from one form to another

Pre-16 – under the age of 16

Printed circuit construction – a method of constructing an electronic circuit in which the electrical connections and contacts are made using copper tracks and pads that are 'printed' on the surface of an insulating board. Simple circuit boards use 'through-hole' mounting of components before their leads are soldered and trimmed (where necessary)

Process – a systematic series of actions

Process methods – how it will be made

Product – a component or complete item produced within a sector, or a service carried out

Production plan – a detailed strategy for producing an engineered product or delivering an engineering service that describes the sequence of operations, processes, materials and resources used.

Quality – tolerances in dimensions and other essential features

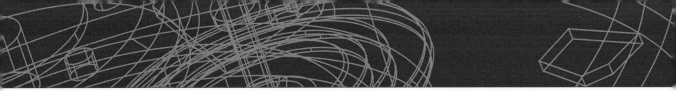

Quality control – activities or techniques to check a product or service is fit for purpose

Quantity – how many of each product and parts of the product to make

Quick-release disposable tips – usually made from carbides or other ceramics and often golden in colour, can be supplied in different shapes and can have more than one cutting edge to allow them to be rotated when needed. Removal and replacement is easy and quick

Recognition – identifying the type of object present

Recycling – reuse of a product when no longer needed. Recycling usually involves extraction of materials such as metals, plastics and glass so that they can be re-used in the manufacture of new (and possibly different) products

Reject – item of inferior quality discarded after checks Reorder stock level – the number of parts held in stock that trigger the need to reorder a quantity of parts from the supplier equal to the maximum stock level

Repayment period – the time that it takes to repay a loan Research – detailed study or investigation designed to expand understanding of a scientific principle or concept

Resistance – the opposition to flow of current

Resistor – a component that introduces resistance into a circuit causing current flow to be

Responsibility – being answerable, in the first place to the other team members

Robot – a machine that functions in place of a person or persons

Roles – specified tasks or functions which would normally be divided up within the team

Safe product – a product that represents minimal risk and offers consumers a high level of protection

Safety stock level – the additional number of parts to be held in stock to cope with worst case conditions

SatNav – satellite navigational system exploiting GPS (global positioning system)

Schematic – a simplified diagram showing how various component parts are connected together and relate to one another

Sector – a grouped engineering skill or occupation

Self-employed – anyone who works for gain or reward other than under a contract of employment

Shape or form – the overall three-dimensional representation and what the product looks like

Size – the dimensions of each part, position of features, etc.

Social and economic development – how people's lives change, including their wealth

Static electricity – an electronic charge that can build up due to friction between two surfaces

Statutory – this means that something is binding in law and it is a criminal act not to obey it

Stock level – the number of spare parts held in stock and immediately available

Surface finish – a coating or other finish applied to an engineered product to improve appearance, durability or corrosion resistance

Surface mounting - a technique used in the automated assembly of electronic circuits. Surface-mounted components do not have conventional leads and are soldered into place using a solder paste that is applied to their mounting pads on the surface of the printed circuit board

Surface texture – those irregularities with spacing that tend to form a pattern or texture on the surface; this texture has roughness and waviness

Sustainable – property of a product that uses raw materials that can be regenerated or re-grown. Wood is a potentially sustainable material, because forests can be replanted when trees are cut down to produce paper and building materials. Petroleum based products are unsustainable products because they will eventually run out

Team – people working together, but more than just a group

Technical specification – a set of key performance data for a product

Test point – a point in a circuit where test voltages and/or currents are taken

Thermoplastic – polymer material that softens when heated and hardens to a rigid state when cooled. Further heating will cause the material to return to its softened state

Thermosetting plastic – polymer material that can be shaped and cured into a harder, more rigid form, through heat or chemical reaction. It cannot subsequently be softened by heating

Tolerance – an allowable deviation from the desired size (no size can be achieved exactly)

Tool offsets – when tools are mounted, the amount of extension of different tools from the tool holder varies according to their size and length. Tool offsets are, in general terms, these differences

Torque – a force that causes turning. The amount of torque (in newton-metres) is equivalent to the product of the force (in newtons) and the radius at which it is applied (in metres)

Transformer – couples a.c. power from one circuit to another; can step-up or step-down voltage

Transistor – a semiconductor device that can be used as an amplifier, switch or power controller

TTL – a logic family based on transistor-transistor logic

Turning – material removal process typically generating/forming shapes cylindrically with the work rotating

Variables chart – measures a variable characteristic along a scale, such as length

Voltage table – normal working voltages for a circuit, often included in service information from manufacturers.

Wire frame image – drawing in a CAD or simulation package showing the outline of the image, so that it can be looked through and show internal elements

Work piece datum – a reference point that measurements are taken from, which may be negative

Work-holding – device or devices that position work to meet the required cutting motion

Work instruction/job sheet – a description of a particular operation or task that specifies what should be done, how it should be done and what materials, processes and tools should be used

WRL – Work-related learning, which implies that the learning is achieved through an employment style situation

Engineering
Level 2 Higher Diploma

Principal Learning

Edexcel's own learner resources for the Diploma

Bring your learning to life with this engaging student book, packed with essential information, real-life examples and motivating activities.

Each topic is covered in a double-page spread, so everything you need to know is manageable and easy to digest.

Supporting your learning through:

- **Motivating real-world case studies** and practical activities
- **Personal learning and thinking skills** and **functional skills** integrated into activities, giving you plenty of practice
- Tips on **how to manage your project** alongside your learning – and get the best out of both!
- **Assessment advice** so you can aim for higher grades

About the authors

Our author team combines wide-ranging industry experience with in-depth knowledge of the Diploma:

Mike Tooley, a respected engineering textbook author and consultant, is former Vice Principal at Brooklands College, Surrey and a contributor to the Edexcel Diploma specification.

Mike Deacon, consultant, Chief Examiner and Diploma specification developer, has wide experience working in both industry and Further Education.

Nik O'Dwyer is Head of Engineering at Leiston High School, Suffolk, a specialist technology centre.

Level 1 Foundation Diploma: Resources

Engineering
Level 1 Foundation Diploma
Student Book
978 0 435756 25 3

Engineering
Level 1 Foundation Diploma
Teacher Resource Disk
978 0 435756 26 0

Level 2 Higher Diploma: Resources

Engineering
Level 2 Higher Diploma
Assessment and Delivery
Resource with CD-ROM
978 0 435756 21 5

Distributed by Heinemann on behalf of

01865 888 118

ISBN 978-0-435756-20-8

9 780435 756208